Lucius Daniel Davis

Ornamental shrubs for garden, lawn and park planting

With an account of the origin, capabilities and adaptations of the numerous species

and varieties

Lucius Daniel Davis

Ornamental shrubs for garden, lawn and park planting
With an account of the origin, capabilities and adaptations of the numerous species and varieties

ISBN/EAN: 9783337224042

Printed in Europe, USA, Canada, Australia, Japan

Cover: Foto ©berggeist007 / pixelio.de

More available books at **www.hansebooks.com**

ORNAMENTAL SHRUBS

FOR

GARDEN, LAWN, AND PARK PLANTING

WITH AN ACCOUNT OF THE ORIGIN, CAPABILITIES, AND ADAPTATIONS OF THE NUMEROUS SPECIES AND VARIETIES, NATIVE AND FOREIGN, AND ESPECIALLY OF THE NEW AND RARE SORTS, SUITED TO CULTIVATION IN THE UNITED STATES

BY

LUCIUS D. DAVIS

FULLY ILLUSTRATED

G. P. PUTNAM'S SONS
NEW YORK AND LONDON
The Knickerbocker Press
1899

PREFACE

AS this is not designed to be a scientific treatise, no attempt is made at strictly botanical classification or description. What is written is more especially for the large number of people who, though interested in plants and flowers, have little or no knowledge of botany, and neither time nor inclination to acquire it. It is not intended by this statement, however, to give support to a somewhat common opinion that the lessons of botany are useless or uninviting, for few studies can be, to a genuine lover of nature, more attractive or even fascinating. In nearly all cases the popular names of plants are given in connection with those by which they are scientifically known throughout the civilized world. Botanical terms and phrases are employed in description only when it is believed they will interest and assist the ordinary reader, rather than tend to his embarrassment. It is certainly worth something to those who admire trees, shrubs, and flowers to know their scientific as well as their common names, and, to some extent, their origin and history.

It is as a help to such knowledge, the want of which is sorely felt by many, that these pages have been written and are now given to the public. Much that is contained herein is gathered from the writings of those who have gone before, including recognized authorities whose works are valuable chiefly to those who, like their authors, are learned in botanical studies. But dependence has not been made on these alone. The volume has been prepared in Newport, R. I., America's great summer resort, which in its magnificent villa and cottage grounds is almost literally a city of gardens and flowers. Here perhaps more than anywhere else in America are to be found in practical use the combined horticultural treasures of the world.

These famous gardens derive their chief beauty and glory from what are known as hardy plants. In almost every instance the chief reliance for both flowers and foliage is upon shrubs and herbaceous perennials. The author has improved the opportunity of studying the processes of growth and cultivation on most of these estates from their inception to their present proportions, and is thus able to write largely from personal observation and study of the living specimens in all stages of their growth. Here are to be found the newer as well as the older hardy exotics from all parts of the world, where such have been grown alongside our native plants and their relative merits fully determined. All those which have withstood the tests of experience are here brought under review, and their especial characteristics noted so far as practicable in the space allotted; it being the purpose to cover the whole field especially of the hardy shrubs, old or new, adapted to useful and ornamental planting.

It is well understood that botany deals chiefly with fixed forms, as represented by orders, genera, and species,

Preface.

and that it takes little or no note of such varieties as are constantly making their appearance throughout the world. For this there is good reason from a scientific standpoint, but when it comes to the practical use of plants in general cultivation it is found that many of the species thus treated have given forth varieties, through processes well understood, that are far more valuable for the work in hand than the originals, and such as are coming, in a large measure, to displace them. A very large proportion of the plants in the best gardens of Europe and America belong to the latter class, many of which are not even named by the scientists—much less described. To these especial attention is given, as for horticultural purposes they are of great value. It is true that much has been written in a fragmentary way concerning these varietal forms, but this is believed to be the first attempt to gather and publish in a single volume an account of the wonderful evolutions in connection with the several types so far as they are of practical use in our gardens and parks. There are also many excellent books in the hands of the people, or at their command, treating of the plants of certain sections of the world, each complete in itself, but regardless of their value in horticulture or of the uses to which they may be put. All this is in the direct line of scientific inquiry, and such books are of the highest possible value, but fail to meet the call for information which comes from the man with grounds to plant, and who is neither a botanist nor versed in horticulture.

Though the attempt is here made to describe in brief the desirable forms indigenous to other countries as well as our own, so far as they are in use among us and applicable to the wants of American horticulture, there will still be left large possibilities for the future. New varieties are springing up and new forms appearing every year, both by natural processes and through the skilful work of the hybridizers and gardeners, who are ever on the lookout for new things in this line. These processes will go on in the future as they have in the past, and it may well be believed that the possibilities are limitless in this direction.

Acknowledgments are due to Messrs. Elwanger & Barry of Rochester, N. Y.; The Gardening Company, and the publishers of *Park and Cemetery* of Chicago; E. L. Beard of Boston, and W. C. Egan of Highland Park, for several valuable plates and photographs used in illustrating this volume.

L. D. D.

NEWPORT, R. I.,
February, 1899.

ORNAMENTAL SHRUBS.

KALMIA—Mountain Laurel.

THE kalmias, or laurels, are among the most beautiful plants in cultivation. They constitute a small genus of the order *Ericaceæ*, which was named by Linnæus in honor of Peter Kalm, who was at one time his favorite pupil, and, later, a traveller and distinguished botanist. They are all of American origin, and may be found over a large extent of territory ranging from Canada to Florida. But five or six species are known, and not many established varieties, though there are some forms so near the border between the two as to make it difficult to draw the line with certainty.

K. latifolia—Calico Bush.—This is the well-known mountain laurel, which is indigenous to New England and even much farther north, and may therefore be put down as perfectly hardy and easily grown throughout the north and northwestern States and Territories. It is found high up among the New Hampshire mountains and often in most inhospitable situations, as well as upon the Alleghany ranges and Georgia hillsides. In protected situations, it sometimes grows to a height of fifteen to twenty feet, but as usually seen, its proportions are very much less. It has a slender stem, with branches in twos or threes in imperfect whorls. The leaves are scattered, though often in tufts, from two to four inches long, rather narrow, acute at each extremity, glossy green, coriaceous, and continuing during the winter even in the coldest climates. Few or no plants produce more lovely blossoms, which appear in June and July, and in thus following the rhododendrons and most of the azaleas, are of the most effective service in keeping up a succession. They are in terminal heads on flower stalks an inch or more long. The color of the corolla varies from a pure white to a rich rose, with numerous shadings between the two. The border of the tube is painted with a waving, rosy line, and the pencilling is as delicate as can well be conceived. The wonder is that a shrub of so great hardiness and such charming flower is not planted much more largely than it is. It is more easily grown than the rhododendron and is in nowise less desirable. It is suggested that the difficulty experienced in removing plants from the woods to private grounds, and the many failures in that direction, have created the impression that it is unusually fickle and cannot be depended upon. But, as a matter of fact, it is no more so

BROAD-LEAVED LAUREL.
(KALMIA LATIFOLIA.)

than numerous others of the best and most common plants in our gardens. If one will go to the nurseryman instead of the woods, he will find very little difficulty in this direction. Kalmias properly grown and trained yield as kindly to removal as do most other plants, and can be handled as safely. Nicholson pronounces this "one of the most useful, elegant, and attractive of dwarf flowering shrubs."

K. angustifolia, or narrow-leaved laurel, is a low evergreen plant, usually from one to three feet high, and is often found growing in bunches or paths in moist or low grounds, where it is deemed especially undesirable by the farmer or herdsman who considers it poisonous to calves or lambs. So common is this impression that in many sections it is known as the lamb-kill or sheep-kill plant. It is claimed by some good authorities that the foliage is not poisonous at all, and that the ill effects ascribed to it come from the fact that the foliage is quite indigestible, and thus fatal at times to young and tender animals. This is all the more probable from the fact that we seldom or never hear complaints of fatalities in the case of cattle or sheep of mature years, which, it is to be presumed, feed on the leaves as freely as do their young. In this little shrub the flowers are in lateral corymbs, and in from three to twelve whorls to each spike. They are purple and crimson, and appear in early summer. The London *Garden* says that *K. angustifolia* should always be planted in rhododendron beds so as to keep up a succession of flowers, and mentions three sorts which may well be used for such a purpose. There are several pretty varieties, one of which, the *nana*, makes an excellent pot plant. It grows but six to eight inches high. *K. glauca* is another dwarf of from one to two feet, having lilac-purple flowers, and leaves with revolute edges, long and narrow, green on the upper side and glaucous white beneath. It is a handsome little shrub, and can be used to advantage in many situations. *K. hirsuta* is an extreme southern species, ranging from South Virginia to Florida, and is said to be found also on the island of Cuba. The flowers are rose-colored, and appear later than the others. It is not of much worth, however, for garden purposes.

DEUTZIA.

THE deutzias constitute a genus of the order *Saxifragcæ*, and are mostly natives of Japan and the Himalaya Mountains, though it is believed that they are also indigenous to northern China and perhaps other portions of Asia. None of the species is found in Europe or America as native of the soil. They received their name in honor of Johann Deutz, a Dutch naturalist, whose memory as a botanist is thereby carried to succeeding generations. Nearly all are hardy shrubs, with rough bark, axillary flowers, and leaves mostly ovate, acuminate, serrate, and more or less scabrous. Though hardy, some of the smaller members of the group are suited to forcing under glass, by which process they are made to produce beautiful flowers at any season of the year desired, and in great abundance. When introduced to Europe they were received with much favor and were soon widely distributed, as was also the case in our own country, where they still occupy an important place in garden and park

Deutzia. 5

planting. The genus is not large, but the number of varieties is constantly increasing, some of which are in marked distinction from the type.

D. crenata is now described as the type from which several others, heretofore classed as distinct species, are recognized simply as varieties. It is a fine shrub six to eight feet high, and often throwing up several stems from the same root, the whole forming a well-proportioned head quite as broad as its height. The leaves are ovate-lanceolate, serrulate, somewhat rigid or stiff, and rough to the touch. The flowers are white, in racemes or panicles, and very pretty. It was at one time largely planted, but in later years has given way to some of its varieties which have been found to possess all its good qualities with some others in addition. *D. c. candidissima plena* is one of these, of which scarcely too much can be said in praise. The white blossoms are double and so numerous that the bush in its flowering season has the appearance of a mass of small rosettes. *D. scabra* has long been spoken of as a species, but is now counted as another variety of the *crenata*. It, too, is a good plant, having single flowers, white within, and marked with pink or purple on the outside of the calix. *D. waterii* has also been claimed as

DEUTZIA CRENATA.

a species, but is now generally held to be a varietal form. The flowers are double, pinkish-white, opening nearly flat like a rose, and of larger size than in most of the other forms. As it is still rare the full value of the shrub in its adaptations to various localities is not yet determined, but the promise is good. All these forms are hardy as well as desirable.

D. gracilis is one of the smallest members of the family, and is widely known in cultivation. It usually grows from two to four feet, with numerous slender branches, which combine in the formation of a symmetrical and well-rounded head. The flowers are small, pure white, and produced in the most luxurious abundance, ranging along the whole length of the stems, and giving the low bush much the appearance of a large bouquet. They appear in May or early June, leading in this respect most members of the family. There are few plants better adapted to forcing in pots, under glass, or even in a well-warmed and light cellar. For growing in small grounds or fitting into vacancies among larger specimens these low shrubs serve a most valuable purpose. They require but little space, and need only to be cut back to preserve a well-balanced head. The plant is a native of Japan.

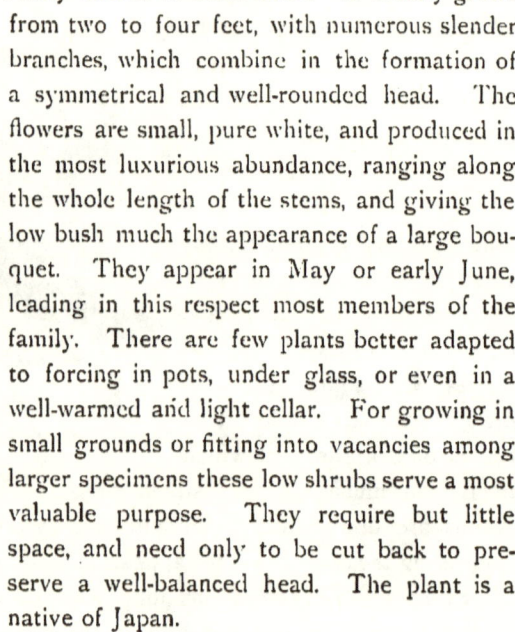
DEUTZIA GRACILIS.

D. parviflora.—This is as yet so little known as to be still classed among the novelties. It is a native of northern China, and was carried from the valley of the Amoor to the Imperial Botanic Garden of St. Petersburg, from which point it has, within a few years, been distributed throughout Europe and America. It grows in clumps, consisting of numerous erect stems or branches from four to six feet in length, which are clothed with dark green leaves, lanceolate, toothed, somewhat wrinkled, and of good substance. The flowers are creamy-white, composed of five petals, and without any splashes of color, the bunches somewhat resembling in size and arrangement those of the lilac. They appear a week or two before those of *D. gracilis*, which has heretofore been supposed to be the earliest as well as most floriferous of all the deutzias, and about a month in advance of most other varieties.

D. lemoinei is a hybrid between the *gracilis* and the *parviflora*, and was brought out by Monsieur Lemoine, the noted hybridist, who has done so much to add to the pleasures of horticulture and the brilliancy of our gardens. The plant is described as having stouter and more upright branches than the *gracilis*, and shorter and more numerous than those of the *parviflora*. The blossoms are about three quarters of an inch in diameter, and are borne in loose, many-flowered terminal panicles on axillary leafy shoots, with pure white, broadly ovate-rounded, spreading petals and reddish-yellow stamens. This is believed to be an improvement on the almost universally popular *D. gracilis*, and destined to largely supersede it as it becomes better known.

D. discolor, var. *purpurascens*, is, perhaps, the latest hopeful introduction among the deutzias to our country.

According to *Garden and Forest* the seeds of this plant were sent to the museum in Paris in 1888, by a French missionary who had discovered it in the Chinese province of Yun-nan. A specimen was secured by the Arnold

DEUTZIA—PRIDE OF ROCHESTER.

Arboretum at Cambridge, where it has flowered two or three years. It is described as " a shrub of neat, compact habit, two or three feet tall, with slender stems, thin, ovate leaves scabrous on the upper surface, and compact panicles of pale pink flowers." It is not yet known that it will endure our northern winters. The place of its nativity

Clethra—Sweet Pepper-Bush. 9

has a semi-tropical climate, and whether suited to New England and the Northwest or not, there appears to be little doubt that it will thrive in the southern and southwestern States, and prove a valuable acquisition.

One of the best of the older varieties, known as the Pride of Rochester, was originated and sent out by the well-known firm of nurserymen, Messrs. Ellwanger and Barry, as an offshoot of *D. crenata flore pleno*. Though but a comparatively recent introduction, its merits are such that it has already become widely and highly appreciated. It carries large, double, white flowers, some parts of the petals being slightly tinted with scarlet or rose, and is said to excel all the older sorts in size of flower, length of panicle, profuseness of bloom, and vigorous habit. It comes into flower soon after the *gracilis* and a week or two in advance of most of the other forms of deutzia.

CLETHRA—Sweet Pepper-Bush.

THE clethras are highly ornamental shrubs, though until quite recently they have not received the attention which their merits demand. So far as known, they are all American plants, and most of them suitable for use in our best gardens, where they are now becoming better known and more largely planted than heretofore. The species capable of the best service in the northern States and Canada is *C. alnifolia*, a small shrub two to four feet high, but in cultivation sometimes reaching double those proportions. It is often found in masses, growing in low or wet places, or along the banks of streams, and sometimes in swamps, where the roots are submerged in early spring or after heavy rains. But it is known that the plant does equally well, even if not better, in garden soils such as are suitable to the rhododendrons and other peat-loving plants. It grows with a slender,

CLETHRA ALNIFOLIA.

straight stem, not much branched, the bark being at first light green and downy, but afterward becoming a dark purple and often striped with gray. The leaves are inversely egg-shaped and slightly pubescent, while the blossoms are borne in terminal racemes and from the axils of the upper leaves and side shoots. They are large, white, and very showy, and when present in masses never fail to

Clethra—Sweet Pepper-Bush.

attract attention. The racemes, which stand erect above the bright, glossy leaves, begin to open in July and continue until October, thus supplying the dullest period of the summer, so far as hardy shrubs are concerned, with abundant blossoms of the highest value. They are withal exceedingly sweet-scented, giving forth an odor not at all disagreeable, but such as makes the plant popularly known as the sweet pepper-bush. It does well in half-shady situations, and cannot be too highly praised for use in landscape work, whether grown in masses or as single specimens.

C. acuminata is also a native of the United States, and is often seen growing on the mountains of North Carolina and other similar locations, where it usually appears as a small shrub of from four to six feet. It is not much known to cultivation in this country, though planted in English gardens and on the continent, where it is quite a favorite. It is there spoken of as a plant growing in tree form, and from ten to fifteen feet high. The leaves are more oval than those of the preceding, and more sharply pointed, having a bluish cast above and being slightly glaucous beneath. The flower spikes are large and conspicuous, though not superior to those of *alnifolia*. *C. paniculata* is also a good plant, but in nowise superior to those already described, the chief distinction being in the form of the flowers, which are gathered in panicles not quite so compact, and slight differences in the shape of the leaves. *C. arborea* is a more tender species, and has been long grown in English greenhouses, though it usually thrives when planted outside, if afforded ample protection. As it, too, is indigenous to the Carolinas, it is altogether probable that it would be a popular open-garden plant farther south, and it is possibly already more or less in use in that section. It is said it has the disadvantage of requiring considerable age before coming to perfection as a flowering plant. But, like all the others, it blooms nearly all summer. There are other tender sorts, some of which are prized for cultivation under glass, but none superior to those already named.

CALYCANTHUS.

THE members of the calycanthus family constitute a small genus of North American plants, mostly confined to the United States, where in their native haunts they are popularly known as Carolina allspice from the fragrance of their flowers as well as of their foliage. Indeed the whole shrub emits a spicy perfume somewhat resembling camphor, including the stem and more especially the smaller branches when bruised or broken. They are more frequently found along the shady banks of streams where there is plenty of moisture, and in situations protected from severe winds, but prove sufficiently robust to maintain themselves in all parts of the temperate zones; though varying in size and attractiveness according to the positions occupied. They are all interesting plants and worthy a place in every considerable collection of shrubs and trees. As under-shrubs they do good service whether planted singly or in masses.

C. floridus.—This is the longest- and best-known species and was described by Loudon in his copious notes on the

Calycanthus.

American sylva. It was introduced to English gardens as early as 1826, where it has since held its own and is still a favorite. The plant forms a small compact bush four to six feet in height, though it is occasionally much taller. The foliage is composed of oblong-shaped leaves, deep green, inclined to be coriaceous, and slightly downy. The blossoms are deep blue shading to purple, one and a half inches across, with petals somewhat fleshy. They appear early in spring, remain well into summer, and are quite numerous. Their long continuance is very much in their favor. Taken all in all, this calycanthus may be accepted as one of the best of our native shrubs.

SWEET-SCENTED SHRUB.
(CALYCANTHUS FLORIDUS.)

C. glaucus is, it may be, a less valuable plant, but is possessed of some interesting features that are worthy of notice. It is not so strongly impregnated with the peculiar odor referred to when its leaves or branches are bruised or crushed, but it is still fragrant to a remarkable degree. The flowers are much the same, lurid blue, and of equally long continuance. The leaves are longer, more sharply pointed, and with more marked pubescence. *C. lævigatus* is found growing freely on some of the Pennsylvania mountains, with taper-pointed leaves, bright green and glabrous. The flowers are intense purple and quite showy. Each of these has given off varieties more or less distinct, but scarcely of increased value.

C. occidentalis is a native of the Pacific coast, and more especially of California, where it grows to a greater size than either of the preceding and is often found nearly or quite twelve feet high. It is there known as the sweet-scented shrub, as in fragrance it is much the same as *C. floridus*. The foliage is composed of larger leaves, and the flowers are also of greater proportions, being some three inches in diameter and of a deep crimson color. It proves the most showy of all the species and a most desirable plant for garden use.

EXOCHORDA—Pearl Bush.

THIS is a genus of but a few species belonging to the *Rosaceæ* and closely related to the spiræas. It is a native of China and has long been known, though not brought into general cultivation until more recently. Its popular name comes from the fact that it is a free bloomer, the flowers being pearly-white, and covering the entire bush. It endures the New England climate, though in the North it usually grows but eight or ten feet, while in the South it often becomes a shrub or tree of twice these proportions. Most of the specimens to be found in parks and gardens do not appear at their best, except in the flowering season, as they are permitted to grow without proper pruning. The exochorda is not seen to advantage when out of bloom, unless it is kept in the form of a compact bush. Left to itself

EXOCHORDA GRANDIFLORA.

it inclines to naked stems and branches and such as are by no means graceful. In fact, as ordinarily grown, its beauty consists only in the numerous large white flowers in April or May; and these are not of long continuance. It has certainly been overpraised in some of the catalogues, but is well worthy a place in the border or shrubbery. It has been suggested that lower and more bushy plants be grown in front and around it to hide its faults, while securing the full benefit of its blossoms, which alone make it worthy of planting.

SAMBUCUS—The Elder.

THE elders are closely related to the viburnums and honeysuckles, which are among the most useful and attractive ornamental plants. There are not far from twenty species belonging to the family, and some of them have done and are still doing good service to mankind as useful plants, while a few, especially of the varieties, are exceedingly beautiful in leaf and blossom.

S. nigra, or black elder, is a native of Britain, and is found growing freely all over the continent, where its fruit has long been much used in the manufacture of wines and the preparation of medicines, and sometimes as an article of food. The regard which was had for this shrub was well expressed by Evelyn when he wrote: "If the medicinal properties of the leaves, bark, berries, etc., were thoroughly known, I cannot tell what our countrymen would ail for which he might not fetch a remedy from every hedge, either for sickness or wound." This high estimate of its virtues may not have continued to our own times, but the good qualities are still recognized in many directions. In extreme cases the European elder grows to a height of from twenty to twenty-five feet, with a well-rounded, bushy head half as broad. The flowers are small, white, and in flat cymes five or six inches across, followed by small, black, berry-like fruit, in great abundance.

There are several varieties of this species which are especially ornamental in European as well as in American gardens. One of these, *S. n. aurea,* golden elder, is one of the very best yellow-foliage plants in use for decorative purposes. The color is solid and far more permanent than with many others which start out well and then fade away. For best effects it must occupy a sunny position, and be well pinched back, so as to compel a dwarfish habit. Thus planted and maintained, when grown in masses it is unexcelled. Another sort, *S. n. laciniata,* or parsley-leaved elder, has its leaflets curiously and finely cut into segments, which retain their natural color, and produce a good effect. It, too, is a fine shrub for massing or edging. *S. n. variegata* has its foliage in the typical form, but marked with white, the contrasts being so sharp as to render the plant a decided curiosity as well as a thing of beauty.

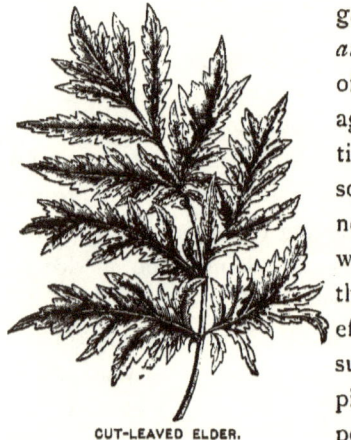

CUT-LEAVED ELDER.

Sambucus—The Elder.

There are several other well-known varieties, but their merits as ornamental shrubs are inferior to those already named.

S. canadensis is the well-known common elder of America, found everywhere from Canada to the Gulf States, growing preferably in moist locations, but making itself quite at home in the fence corners and by the roadside, wherever given a chance to grow. Its pithy stems, well bunched, are from five to ten feet high, having compound leaves with from five to eleven leaflets, mostly smooth and three-parted. The flowers are white, gathered in flat cymes, and succeeded by dark purple or black fruit, which is often used in the manufacture of domestic wine, for which it has especial adaptations, and occasionally for making tarts and pies where more desirable berries are not to be obtained. This species appears to have a wider range southward than most of the members of the tribe. *S. racemosa*, another American species, runs wild over a large extent of country, having red berries instead of purple, but not differing essentially otherwise from the preceding.

The value of the elder as a seaside plant can scarcely be overestimated, both as a nurse tree and because of its own merits as very ornamental. Says an English writer: "Isolated specimens of it may be seen far out on the dreary stretches of ever-shifting sand, and looking as healthy and robust as we find them in their favored locality —a damp, shady wood. There is not much beauty, some will say, about the elder, though I hold a different opinion; but beauty alone, it should be kept in mind, is not what we are at present in quest of; rather a tree or shrub that can stand the first brunt of a sea storm, and by so doing afford shelter to less favored kinds. Whenever a seaside garden is to be formed, lift a few plants of an elder from some waste or common, and plant them—in pure sand, if you like—on the outer margin of the sea screen, and in a short time a capital shelter for other choice things will be formed."

PAULOWNIA.

THE *Paulownia imperialis* is a Japanese tree of striking appearance and with many peculiarities. It was named by Europeans, on becoming acquainted with it, in honor of Anna Paulowna, Princess of the Netherlands and daughter of Paul I., Emperor of Russia. It is said that in its native country it grows to a height of thirty feet, but it is not often seen in such proportions in either Europe or America. When first brought to France the tropical appearance of its foliage led to the conclusion that it must be an exceedingly tender sort, and so in the famous Garden of Plants in Paris it was treated as a greenhouse shrub. But it was not long before it proved itself sufficiently hardy to withstand the winters of that locality. It was then regarded as a great acquisition, and young plants were sold in the English markets at high prices, ranging from three to six guineas each. But these figures did not rule long, for it was soon ascertained that the newcomer was not only hardy, but of the easiest possible propagation. It can be grown not only from seed and from cuttings, but by a division of the

Paulownia. 19

roots, small pieces of which will produce vigorous plants the first season. Even the buds will grow, like those of

PAULOWNIA IMPERIALIS.

the mulberry, if taken off in the spring and planted in a hotbed or under a hand-glass.

Few plants are of more rapid growth, the young shoots

often reaching eight feet the first season. These are covered with immense foliage, the individual leaves being somewhat heart-shaped and a foot and a half broad by two feet long.

Though the paulownia does fairly well in New England and New York, it should be planted there only in protected situations, for its top is too large to withstand severe windstorms. The tree is more especially adapted to a warmer climate, and should be entirely at home in the Southern States. Says an English writer in one of the magazines: "To see the paulownia to perfection in Europe one must go to the sunny South, and I have a pleasing recollection of the magnificent avenues I saw of it in full flower at the end of April in the gardens of the Villa Borghese and the Pincian Hill in Rome, where the climate exactly suits it." It should be borne in mind that it may be grown as a tree or a shrub, as when it is cut back from year to year it sends up numerous vigorous shoots, and as few or as many may be preserved as are desired, and they are never more beautiful than during the first season's growth.

Except for their size, the leaves very much resemble those of the *Catalpa speciosa*, but are of a darker green and better substance. It is for the foliage more than the flower that the tree is prized by those who best know its worth. On older trees the leaves are usually smaller, and so less beautiful. For two reasons the tree to be at its best must be sharply cut back every season—one the preservation of good form, and the other of good foliage. An old and misshapen paulownia with distorted and bare limbs is like a plague spot in any garden, and those who are accustomed to see it in that shape will certainly come to the conclusion that the tree has been and is still much overpraised. There is scarcely room for carrying this pruning process too far; for if the last year's wood is cut to the ground in early spring, new shoots will put forth and soon reach six to ten feet, and thus become a thing of beauty scarcely equalled on the lawn or in the border. When wished, the trunk may be pollarded, and thus the head carried as much higher as desired. The individual flowers are about one and a half to two inches long, violet-purple, with dark spots inside, and in terminal panicles of great size. While not especially beautiful, these never fail to attract attention. They appear in June, and are followed by abundant fruit.

MYRICA.

THOUGH this is not one of the most beautiful of American shrubs, it is for some purposes and in some situations one of the most useful. The family is a large one, though only a few species are known in America. They grow in all sorts of soil and far to the north, often taking possession of the hillsides and sandy plains, and so fully occupying the ground as to be regarded a nuisance, especially if the land is wanted for something else. There is a common saying that the roots extend as deeply into the ground as the stems and branches do into the air. However this may be, they evidently come to stay, and in their tenacity and indifference to situation is found one of their virtues when they come to be utilized as garden plants. The best of these, as well as the most common, is *M. cerifera*. It varies in height from one to four feet, but responds quickly when afforded the advantages of cultivation and the use of fertilizers. The plant may be recognized at the proper season by its bluish waxen fruit, found in the axils of the stems and along the branches, which to some extent affords an article of commerce in the form of a valuable wax. This little shrub when planted along the shore withstands the ocean winds and storms perhaps better than any other plant known in cultivation, and can be made to do good service in establishing plantations by the seaside. It is now coming largely into use for that purpose, as it affords protection to more attractive specimens which may be planted to the leeward. Beginning with a hedge of these myricas, plantations may often be established where without something of this nature the task would be hopeless. Almost any bleak and barren exposure can be covered in this way and become comparatively beautiful. With this line of defence other shrubs and flowers may be introduced and made to thrive where without such protection nothing desirable could be made to grow. Thus the wax-myrtle, in itself unattractive and undesirable, is made of especial service in the planting of exposed estates. In the Royal Palm nurseries located forty miles south of Tampa it is included among the native plants of that section, and described as an evergreen producing "slate-colored berries," and pronounced hardy throughout the entire South. Such being the case, the myricas will doubtless be found of service as nurse trees, or shrubs, on sandy and bleak barrens, against tropical exposures as well as northern blasts.

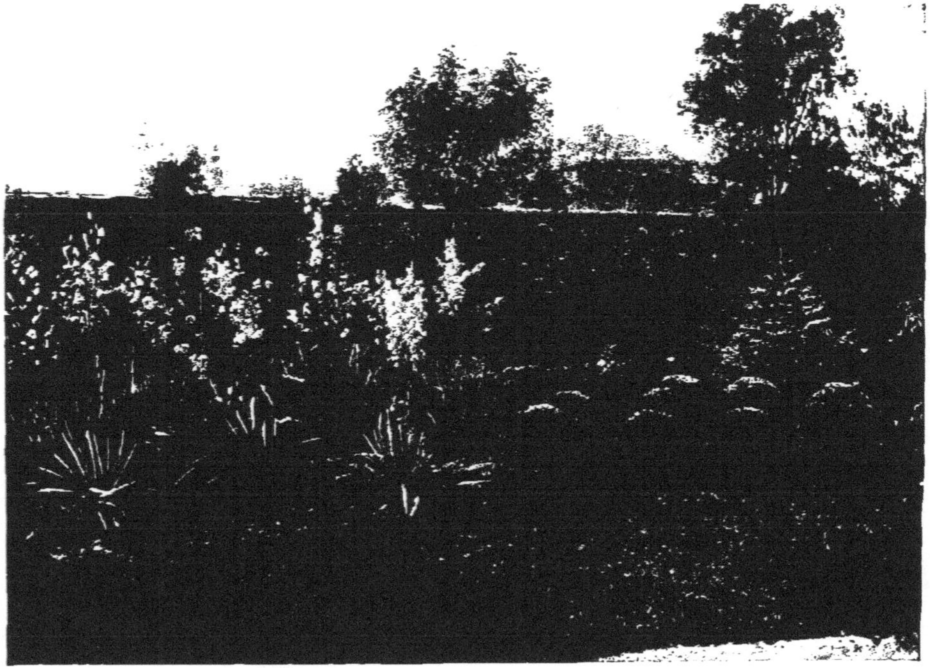

SECTION OF A LAWN AND GARDEN AT NEWPORT.

M. asplenifolia, commonly known as the sweet-fern, also thrives in poor soils and is a good seaside plant. It has fern-like foliage, more attractive than that of the *cerifera,* while its flowers, which are freely produced, are really beautiful. It grows about three feet high and is coming to be planted for its own sake in the ordinary garden. *M. gale* is another form with cut-leaved foliage which is fragrant. It rises about three or four feet and helps cover many a New England hillside with verdure.

AZALEA.

THE azaleas are among the most beautiful and interesting of all our flowering plants. They have been long known in the Old World, and have always been objects of admiration. The genus belongs to the natural order *Ericaceæ,* and in many of its characteristics is allied to the rhododendrons, though mostly having deciduous leaves, and flowers with five stamens instead of ten. There are numerous species and varieties, some hardy and others suited only to hothouse culture.

A. pontica has been, perhaps, longer known to civilization than any other, and it is certainly one of the most attractive of the family. Its native habitat appears to have been in the countries about the Black Sea and along the northern shores of the Mediterranean. In later years it has been carried to all parts of Europe, and largely cultivated wherever the climate will allow. The plants grow from three to five feet, often presenting a broad, round head, with large, oblong, glossy leaves, and masses

Azalea.

of blossoms, with which the branches are so completely covered as to be almost hidden from sight. The flowers are somewhat funnel-shaped, with long stamens, and among the numerous varieties are flowers of many colors. They are very fragrant, appearing in May or early June. Unfortunately, the species is not entirely hardy, and in many situations needs winter protection. Almost every part of the plant is narcotic, and to some degree poisonous. Honey gathered from its flowers is known to produce stupefaction in the bees that gather it, and sometimes death. It is believed to have been such honey that caused the delirium among the soldiers of Xenophon's army, and compelled the famous retreat that has figured so prominently in ancient history.

A. indica is indigenous to the East Indies, and is common to China and Japan. Since its introduction to Europe and America it has been largely grown under glass, as it is too tender for our northern winters, though it may do well in the South. The flowers are mostly solitary, and always very beautiful. Its chief value to us in open-air cultivation has been the impartation of its splendid colors to the varieties produced by crossing with hardier sorts, and there are enough such hybrids to meet all the requirements in this direction.

A. mollis is a somewhat recent introduction from Japan, which has been received with much favor. It proves to be adapted to forcing under glass, but, as it is hardy, its chief cultivation is in the open air. It is of low, spreading growth, so that its diameter is often greater than its height. It is sometimes grafted so as to appear in tree form, thus affording a round, well-shaped head on an upright stem, at such height as may be desired. With the Japanese it is said to be the favorite among azaleas, and is largely grown. It rises to a height of three to four feet, having deciduous leaves, elliptic in form, with ciliated margins, green above and almost silvery-gray beneath. The flowers of the type are campanulate, somewhat downy, and flame-colored. But some of the varieties produced from seedlings and by hybridization furnish gorgeous blossoms of white, yellow, and orange, each in some cases more or less tinted. They appear early and, being large, are very showy. The *A. mollis* needs only to be known to be appreciated as among the best of the class.

AZALEA MOLLIS.

A. nudiflora, known as the wood-honeysuckle, is an excellent little shrub for planting in the border or in groups. It is of American origin, and is quite common in most parts of the United States. In some respects it resembles the *pontica* and its hybrids, but is quite inferior

AZALEA NUDIFLORA.

Azalea.

in brilliancy and tone. Its numerous flowers are large and very showy, presenting a beautiful appearance in early spring. The shrub is taller than the preceding and much more hardy, as it endures the climate as far north as Canada. It grows well in any good soil, and needs but little care. The flowers are light pink, and appear about the middle of May. The shrub grows to a height of six to ten feet.

A. calendulacea is a native of the high mountains of North Carolina and other Southern States, where it often grows in such profusion as to make the mountainsides appear, from a little distance, as though covered with a robe of scarlet. It is of larger growth than most other species, either native or foreign, and has the peculiarity of blossoming late in the summer, after all the others have ceased to show color. It is popularly known, from the gorgeousness of its hues, as the great flame azalea—a name not at all inappropriate, in view of its crowded mass of scarlet blossoms. Many regard it as the most beautiful of all the native American plants, and not wholly without reason. In England and continental Europe it awakened great enthusiasm when first introduced, and it is still regarded as one of the very best of the flowering shrubs.

A. arborescens, or the tree-like azalea, grows from twelve to twenty feet high, and is supposed to be the largest member of the family. It also is of American origin, and produces reddish flowers, in themselves attractive but less brilliant than those of most of the species already mentioned. It is, however, worthy of cultivation in the border, which it greatly helps to enliven with color in early springtime.

A. viscosa is another native species, popularly known as the swamp-honeysuckle, or the pinxter, having highly fragrant, white, clammy flowers. It is too familiar to need detailed description, and, though not equal to many of the others, is in itself an interesting shrub. A variety designated as *A. v. nitida* has its white flowers tinged with red, and is worthy of cultivation. It is smaller than the type, and, like it, is found mostly in swamps, though thriving in any good garden soil.

The so-called Ghent azaleas are probably now in more general cultivation than any others. They are hybrids of the *pontica* and *indica* with *A. calendulacea*, and perhaps other hardy species, and so combine the beauty of the one type with the strength and vigor of the other. This work of crossing has been carried on to such an extent and with such skill that a new class has been established of the most beautiful plants to be found in the gardens of the world. Most of them are perfectly hardy, and are grown in the vicinity of Boston and Chicago, without especial protection from the hot sun in summer or the intense cold of winter. Of course those offered in the market are grafted or budded, and the varieties are so many that it is impracti-

HYBRID AZALEA.

Azalea.

cable to describe them, and the more so as new forms are constantly appearing. It will be sufficient to name some of the most desirable sorts, indicating their colors and such other peculiarities as may be of importance : Admiral de Ruyten, dark rose; Alba Lutea Grandiflora, large, white and yellow; Amabilis, rose-orange; Aurantiaca, orange scarlet; Aurore de Royghem, large, orange and pink; Bicolor, orange-yellow and white; Bouquet de Flore, pink and white; Bronze Unique, dark rose, orange; Comble de Gloire, fine, rose, light orange; Concinna, dark orange; Conspicua Grandiflora, rose-orange; Cruenta, fine, scarlet; Cymodocee, scarlet-crimson; Dr. Gray, Flushing seedling, large, scarlet-orange ; Emilie, splendid dark crimson; Flushing Queen, Flushing seedling, deep salmon; Gloria Mundi, scarlet-orange; Grand Duc de Luxemburg, fine, rose and orange; Jules Cæsar, dark rose and orange; La Superba, rose-orange; L' Intéressante, rose-orange; Macrantha, large, yellow; Mirabilis, rose-pink; Ne Plus Ultra, extra, orange; Othello, splendid, rose; Penicellata Stellata, straw and salmon, late; Plumosa, light pink-orange, early; Punicea, crimson-scarlet; Quadricolor, light rose and yellow; Reine des Pays-Bas, extra, crimson-scarlet; Richardii, light pink and yellow; Rosea Rotundifolia, large, rose-orange; Triomphans, buff, rose and orange; Vandyck, dark crimson and scarlet; Vesuvius, splendid rose-orange; Viscocephalum, white, very fragrant; W. C. Bryant, Flushing seedling, deep pink.

A. amœna.—This beautiful plant heads another group of azaleas, with, as a rule, smaller but not less choice flowers, that are coming into especial prominence. Though not widely known, it is highly esteemed in the best gardens of Newport, where it may be seen in considerable numbers and always in effective combinations. It is a low, bushy shrub from China, from two to four feet high and of a spreading habit, so that its diameter is, or can be made, almost equal to its height. The evergreen foliage is composed of a multitude of small leaves, which become bronzy-lilac or purple in winter, and lose none of their beauty, even under the severest climatic tests. In April or May the bush is covered with masses of rich crimson and purple blossoms, about one to one and a half inches across, and exceedingly attractive. Planted in groups or in rows by the pathway, it is effective the year round, and all the more valuable because of its winter garb. There is a variety known as *Caldwelli*, held to be of freer growth than the original, whose blossoms are nearly twice as large. It is not much known in America, but in England and on the continent is said to be crowding the typical form for the supremacy. Either of these makes excellent borders for rhododendron beds or the larger azaleas, and is an acquisition for the conservatory as well as for garden cultivation.

An English writer through the London *Garden* speaks of some of the newer forms of the azalea, most of which are scarcely known in this country, if at all, and his account of these is made the basis of the descriptions and characterizations which follow. *A. cayciflora* resembles the *amœna* in the hose-in-hose conformation of the flower, but differs totally from it in the color of the blossoms, which are bright salmon-red with a distinct orange shade.

Azalea.

It is an introduction from Japan, a decided acquisition to the azalea family, and certain to be more extensively grown when better known. *A. obtusa* is another of these low-growing forms, without the enlarged segments so noticeable in the case of the preceding. The color is

AZALEA ROSÆFLORA.

much the same, and may be characterized as an orange-red. A variety of this species has blossoms which are pure white, or occasionally striped with red. The flowers are brought forth in great profusion. It is said to have come from China and not to be entirely hardy in exposed situations. *A. rosæflora* is quite distinct from any other azalea, but may not prove to be entirely hardy in northern localities. The flowers are of a pleasing shade of salmon-pink, larger than is usual with this class of shrubs, very double, and imbricated. Its specific name comes from the fact that its buds resemble those of the rose when beginning to expand. For forcing and in southern latitudes it must prove a distinct gain. It is best grown in partial shade. It is known also as *A. balsaminæflora*.

In addition to these there is a numerous progeny of crosses and hybrids, mostly between the *amœna* and some of the *indica* section, and, as might be expected, some of these are of a striking character. In habit they are mostly midway between the two parents, the *amœna* affording the elements of strength and endurance, and the *indica* brilliancy and charm of color. Among the best of the early sorts thus produced and brought to the attention of the public are Mrs. Carmichael, named in honor of the wife of the hybridizer, with purple flowers; Duke of Connaught, dark rose; Princess Beatrice, pale mauve; William Carmichael, carmine suffused with magenta; and Miss Buist, pure white. Later on other experimenters have brought forward the Duchess of Albany, pure white and semi-double; Illuminator, rose-magenta, with vermilion centre; Fosterianum, white and very large, with lemon tint in centre; Hexe, a free-flowering sort and said to be one of the best of all. But it is of little use to continue the list of varieties, which are every year increasing in numbers, though not always in character. Enough have been named to show the possibilities in this direction, and they are certainly full of promise.

FORSYTHIA.

THE forsythias, coming from China and Japan, belong to the order *Oleaceæ*, and have long been in cultivation in English gardens, the name of the genus having been applied in the last century, in honor of William Forsyth, the king's gardener at Kensington for many years. There are but few species or varieties known to us in cultivation, but all that are thus employed prove to be charming plants, and of especial worth because of their season of flowering and the situations which they may be made to occupy to advantage. The flowers are solitary, bright yellow, and very numerous, and so distributed along the branches as to often cover the entire bush. These are produced on the wood of the previous year's growth, and it sometimes occurs that people who do not appreciate this fact prune their plants in winter, thus removing nearly all the flowering wood, and then complain that their forsythias do not meet their anticipations. These plants need to be severely cut back each year, but it should be done at the close of their flowering season, which is before the foliage fully puts out rather than after. Nearly all the wood of the year preceding should be cut away, and the knife may be used without fear of doing harm, as new branches will quickly take the places of those removed. All the forsythias are rapid growers, and the long, slender branches, newly formed, carry foliage sufficiently attractive to justify the highest expectations in seeking a desirable plant of its proportions. When occasion requires, the shrub may be trained on a wall or fence and made to cover a considerable space. It is equally fitted to be formed into a round, compact head, as is often done in the best gardens.

F. veridissima takes its specific name from the bright green leaves which it carries, rather than from the color of its flowers, which are golden-yellow and among the first to appear in spring. This is the plant longest and best known in our gardens, and the species which drew so many praises from flower-lovers a hundred years ago and which were by no means unmerited. It is of erect, spreading habit, and entirely hardy. *F. suspensa* differs but little from the other form, except that its long, slender branches are slightly pendant at their terminals, and so are by many esteemed more graceful. The blossoms may not be quite so numerous, but the plant as a whole is fully as good, though not especially to be preferred except for training on walls or trellises where it will have a wider spread. A correspondent in *Meehan's Monthly* gave an account, some time since, of an instance where a plant was kept to a single stem for ten feet, and then allowed to spread on the trellis, where it did good service by way of affording shade to a doorway which was too sunny an exposure. This illustrates its capabilities in that direction. The plant is sometimes catalogued under the names *F. fortunei* and *F. sieboldi*, but these are to be regarded merely as synonyms.

WEEPING GOLDEN BELL.
(FORSYTHIA SUSPENSA.)

Desfontainea.

DIRCA—Leatherwood.

THE popular name for the *Dirca palustris* is leatherwood. This usually grows in the form of a small tree and to a height of but from three to five feet. It is a native of America and though originally found in moist, peaty soils, can be successfully cultivated under ordinary conditions. The flowers appear early in the season in advance of the foliage. They are yellow with a greenish cast, in terminal bunches, and quite pretty. The leaves are lanceolate in shape, and of a yellowish-green color. The leatherwood is especially interesting from the peculiarities of the thick, porous bark, which is so soft as to yield to the touch like so much putty, though resuming its shape when the pressure is removed. But, while being thus pliant and porous, it has such great strength that a strip half an inch in width is said to be too much for an ordinary man to break. It was formerly used by the Indians for strings to their bows, and for fish-lines. The miniature tree is more valuable as a curiosity than anything else.

DESFONTAINEA.

THERE is but one species known in cultivation, *D. spinosa*, which is a native of South America, having been first discovered on the mountains of Chili, whence it was carried to Europe and received with considerable favor. It belongs to the natural order *Loganiaceæ* and in many respects bears close resemblance to certain members of the holly family. The shrub is a low grower, much branched, and in the milder districts very much at home, though it cannot be safely planted in New England unless especial protection is afforded in winter. The foliage is thick and glossy like that of the English holly, the leaves being entire and armed with spines of considerable prominence. The flowers are at the ends of the branches, solitary, and appearing as late as August. They are bright scarlet, shaded with yellow, and in tubular form, often covering the entire bush, which grows to the height of from two to three feet. As seen in English gardens, Nicholson pronounces it "a very beautiful, hardy evergreen shrub of easy cultivation." It will thrive in any good garden soil, and its presence in the border in winter helps enliven one's home surroundings. Too many of this class of plants cannot well be employed where a favorable outlook is desired in winter as well as summer.

HYDRANGEA.

THE hydrangeas are among our best ornamental shrubs and are widely distributed. They are of the order *Saxifrageæ*, and the genus includes between thirty and forty species, natives of the Himalayas, the island of Java, China, Japan, the United States, and perhaps other countries. Some are evergreens and some deciduous, and nearly all are beautiful and interesting. In our Northern States only a few are sufficiently hardy to be grown freely in the open ground; but farther south the very best sorts can be cultivated in ordinary gardens without difficulty. As a rule the American species are hardier, but not of so good flowers or foliage as some

Hydrangea.

that come from China and Japan, though they seem to be about all that can be desired. Nearly all have large and broadly ovate leaves, pointed and slightly serrated, and would prove valuable plants if grown for their foliage alone. The blossoms are disposed in cymes, corymbs, and panicles, and are distinguished for their size as well as beauty. The forms with sterile flowers, with an enlarged calyx, are to be preferred, and these are almost the only ones in use, their propagation being chiefly from cuttings or by division of roots.

H. hortensia.—This is a species most freely planted in gardens, and popularly known as the changeable hydrangea. It is a native of eastern Asia, and was introduced from China to England late in the last century, where it was mostly cultivated under glass and given special protection, as it was found too tender for that trying climate. There are now said to be a few localities where it can be depended upon outside in ordinary winters with proper care, but it is still looked upon as suited chiefly to conservatory cultivation. In the northern United States it requires much the same treatment, and for the same reason. And so it is mostly seen in tubs or pots as it appears on the lawns or in borders, where it is always showy, and in readiness, when autumn comes, for removal to the pit or cellar. It is thus grown in New England even more freely than in the South, where less care and attention are required in its use. It has large, ovate leaves, acute at both ends, serrated, and of good substance. The flowers are collected in nearly round balls four to six inches through, appearing in July and continuing a month or more. The color is not only variable, but changeable as the season advances. It ranges through mild shades of creamy-white to pink and blue, and is always beautiful. It is coming to be quite the custom, where large plants are standing in the open

HYDRANGEA HORTENSIA.

garden, to prepare and protect them for winter without removal, which in such cases is somewhat difficult. This is done by bending the branches to the ground as nearly as possible, and covering with earth in the form of a mound. Some of the largest and finest plants in Newport have been treated in this way for a succession of years, and it is seldom that losses occur through this process. There

Hydrangea.

are numerous varieties of *H. hortensia* which are worthy of special notice, and some of which are largely in use. Among these is the well-known *H. otaksa*, with opposite, deeply serrated leaves and beautiful rose- or flesh-colored flowers. It is held to be one of the best forms of the whole group. The *Thomas Hogg* has pure white blossoms, very large and compact and very showy for a long time, and is one of the very best. *H. asizia* has variegated foliage, which lends variety in grouping, but supplies no additional flower charms. *H. Empress Eugenie* has good foliage and very large corymbs of blue and pale rose-colored flowers, and should not be overlooked. *H. rosalba* has its blossoms in smaller heads, but they usually cover the whole bush. They are white and pale rose, coming to perfection in advance of most of the others.

H. ramulis coccineis, known as the red-stemmed hydrangea, proves a very valuable acquisition, as it produces large trusses of well-formed blossoms, rich pink or deep rose in color, and in great profusion. It is comparatively new, having originated in western New York within a few years. *H. rosea* is much the same as *Thomas Hogg*, except that its blossoms are red instead of white. *Alba variegata* and *speciosa* are forms with silvery-white marks on the foliage, especially in early spring, when the effect is very striking.

H. vestita, var. *pubescens*.—This is one of the most hardy forms, and is not so well known as it should be. It is a native of northern China and perhaps other Asiatic countries of that latitude. It is but little known in American or English gardens, and is described as a shrub four or five feet high and often from five to eight feet across, the numerous slender branches being clothed with light green, ovate leaves, pointed at both ends, and retaining the color throughout the summer. The flowers, appearing in June, are in cymes five to six inches in diameter. The ray-flowers are numerous, and as they first come out are pure white, but later change to rose or pink, and hold on until late in autumn. It is pronounced by good authorities the most beautiful of all the hydrangeas that are absolutely hardy as far north as New York and New England, and the earliest to blossom in summer.

H. thunbergii is a species from Japan, with blue or rose flowers arranged in clusters with the sterile ones on the outer rim of the cymes and the fertile ones in the centre. It is a small sort of two or three feet in height, and not as hardy as some other sorts. *H. nivea* is distinguished by having nearly white leaves on the under side, and thus affording valuable contrasts when planted in groups. *H. lindleyana* is of Japanese origin, with long leaves and comparatively small heads of bright pink blossoms. *H. stellata flore pleno* is new and rare, with its merits not yet fully tested. There are numerous other varieties of more or less value, in most of which the differences are so slight and unimportant that to describe them would seem a useless task.

H. quercifolia is an American species, a native of the Alleghany Mountains, though not much planted. It is a hardy shrub four to six feet high, and of sturdy growth, with white flowers disposed in the form of panicles rather than the customary cymes. The leaves are about six

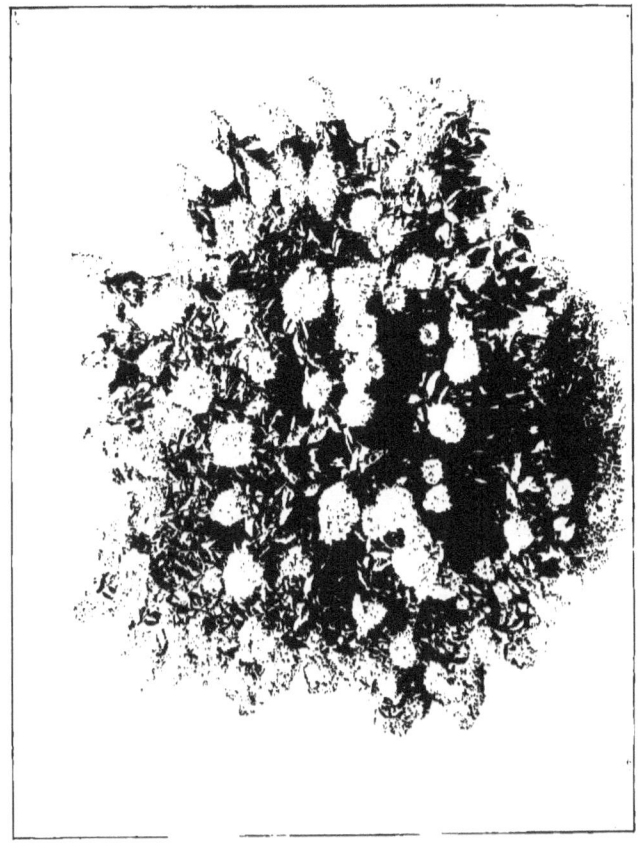

HYDRANGEA PANICULATA.

inches long, obtusely lobed, resembling some of the oaks, and are richly tinted in autumn, giving an especial value in producing good foliage effects.

H. paniculata.—This is probably the most hardy of all the hydrangeas known to cultivation. Its good qualities are intensified in its variety, *H. p. grandiflora*, which has now almost superseded the typical form in our ordinary gardens. It produces immense heads of sterile flowers, often measuring ten to fifteen inches in length and half as many in diameter. These come out in July and continue until frost, being creamy-white at the opening and changing to red later on. The shrub is much branched, and every stem carries one of these immense cones, so that the bush appears as a huge bouquet, arranged with care and precision. If cut back almost to the ground in autumn new shoots arise in considerable numbers, every one of which later during the same season produces a cluster of blossoms. If the shrub is not properly cut in there will soon be too much height, all the last year's growth becoming bare. When desired it can be grown in tree form, with a single stem to the height desired, when a well-rounded head can be established and maintained from year to year. It is often grown in this way, but it is not especially to be preferred. The plant needs no winter protection, but, like all the others, is the better for special care and attention.

H. japonica.—An earlier introduction from Japan, growing about three feet, with ovate, oblong leaves glabrous and finely serrate, with blossoms of bluish cast, though white is predominant, and in crowded cymes. There is a

43 HYDRANGEA PANICULATA GRANDIFLORA—TREE FORM.

variety known as *rosea alba*, which has only the outer individual flowers radiate, and the petals touched with pink or rose.

ACACIA.

THE acacias constitute a large and interesting genus of the natural order *Leguminosæ*. The number of species is estimated at from three to four hundred, and there are in addition numerous varieties and hybrids of especial value in cultivation. They are mostly natives of tropical or semi-tropical climates, and their products are highly esteemed in commerce and medicine. Some of the species furnish the exudation known as gum arabic, some gum senegal, and others a delicious perfume known in trade. Various drugs and medicines are prepared in whole or part from the roots, leaves, and bark, and the bark contains so much tannin as to be also largely used for tanning leather. In some cases the pods furnish material to the native populations for making snuff, and also for distilling an intoxicant which is said to be of a most agreeable nature. They are nearly all beautiful, small trees, with gracefully waving, feathery foliage and abundant, fragrant, pea-shaped blossoms of white, pink, yellow, and purple.

As might be expected, most of the acacias known to colder sections are grown under glass, and very few are entirely hardy in this country as far north as New York and New England. But in the Southern States there are several species well fitted to the climate, and needing no protection; and some of these, with no more care than is often afforded other semi-hardy plants, can be grown as far north as Philadelphia and St. Louis. The leaves are compound, twice and thrice pinnate, the small leaflets being so finely divided as to present a fern-like or feathery appearance. But in some of the species the true leaves are seldom developed. To compensate for their absence, the leaf-stalk, which at first is more or less cylindrical and of small dimensions, becomes flattened and assumes a leaf-like appearance. Thus these dilated leaf-stalks are made to fulfil the functions of leaves, varying largely in the different species. Another curiosity in the structure of the foliage of this class is that the leaves are so placed that their edges look upward and downward, instead of lying flat, as in most cases. One of the results of this peculiar arrangement is that such trees and shrubs afford very little shade, as the sun's rays are but slightly obstructed, so that other plants of suitable size may grow freely beneath. Of the specimens that may be planted safely in the United States the following may be named as among the best:

A. decurrens—Black Wattle.—This is a small tree of good proportions, and highly prized wherever known. It is, withal, of considerable commercial value in its native country, and thousands of acres are said to have been planted in New Zealand and Australia within a few years for the tannin which the bark affords. The tree is exceedingly floriferous, producing long racemes of bright yellow blossoms that contrast favorably with the pale green, fern-like leaves. It was introduced into California some years since, where it was received with much favor. Since then it has been planted in some of the Southern States, and may be grown as far north as Washington under favorable conditions.

A. hispida—Rose-Acacia.—This is a native of the southern Alleghanies, and is one of the most hardy of all the true acacias, and so is more largely known in American gardens than any other. It is a small tree, blooming early in spring, when it is covered with soft rose-colored but scentless flowers. It is a vigorous grower, coming early to maturity, and in every respect is desirable. But even when found hardy it needs to be planted in a sheltered position, as its wood is brittle and the limbs are easily broken by high winds. In England it is often grafted for the sake of gaining height, and as thus grown is very fine.

A. baileyana.—This is a native of Australia, and has been but recently introduced to English and American grounds. Mr. Watson of the Cambridge Botanic Garden, in a letter written from London to *Garden and Forest* some time since, says that he found it growing at Kew, and that it does well in that location. It may not have been tested very thoroughly in America as yet, but as grown at Cambridge it is described as of a close, bushy habit, with stiff, twiggy branches, thickly clothed with short, bipinnate, glaucous leaves, with remarkable glandules on the midrib. The flowers are in small-stalked, spherical heads, deep yellow, fragrant, and produced in large panicles on the end of the branches. In its native country it is said to grow to a height of fifteen feet, but has assumed no such proportions, as yet, in the hands of our horticulturists. It cannot be otherwise than an interesting plant, as the gray bark of the stem and larger branches, with the pale green of the foliage and the bright bunches of golden flowers, constitutes a combination not often seen on the same shrub or tree. The blossoms continue long and answer a good purpose as cut flowers.

A. pubescens.—No one will dispute the value of this plant as ornamental, and it is one of the most hardy and easily grown of its class. Good plants are found in perfection as far north as Boston in some of the best gardens. It was reported as long ago as 1891 that specimens were growing in the grounds of Wellesley College, spreading to nearly, if not quite, ten feet in diameter. These had doubtless received especial care and attention, as do many others of our best ornamental shrubs. In the warmer parts of the country they thrive in the open air the year round. They are beautiful in both foliage and flower, the leaves being finely cut and the blossoms exceedingly abundant. These latter are light yellow, and crowd the long, pendulous branches from end to end, always filling the air with a delicious fragrance. They appear very early in spring, and frequently in winter during a temporary suspension of its customary rigors. It is a good plant for house or church decoration, and capable of filling a want in that direction which has been long felt.

A. farnesiana.—Here also is found a small tree, of about the same height and possessed of many of the same good qualities. Just where it originated is a question not fully determined. It abounds in San Domingo, and is found growing wild in some parts of Texas, if not in others

of the Southern and Southwestern States. With care
and slight winter protection it is found to do well in
the Mississippi valley as far north as Cincinnati, and
presumably is adapted to a considerable southern range
throughout the middle Southern States. It has the
feathery foliage of its class, with flowers of golden-yellow
and most deliciously fragrant. These appear in early
spring in great abundance, and continue through the entire summer and well into the autumn. Cases are
reported where trees from twelve to fifteen feet high carry
heads as many or more feet in diameter, and that, too,
when from but six to ten years old.

ACER—Japanese Maples.

THE Japanese are a patient people. However
others may make haste, they are content to wait
when they have an object in view which they
deem worth their while. Their nurserymen, especially,
are given to processes that an American grower of plants
would scorn to adopt, however much he might desire possible results. In proof of this it is necessary only to note
how largely the foreigners are given to the art of dwarfing
plants and trees and growing them into fantastic shapes.
The beautiful little maples, and many others which appear
in this country from time to time, are largely the outcome
of long-continued artificial conditions. At a meeting of
the Horticultural Congress held in Chicago in 1893, Mr.
Henry Izawa, gardener of the Japanese Commission to the
Columbian Exposition, read a paper by request, illustrating this phase of Japanese work; and as it throws

Acer—Japanese Maples. 49

so much light on the general subject, and explains so many of the peculiarities of certain Japanese plants, a brief quotation may not here be out of place.

In setting out to produce these abnormal results, Mr. Izawa says that the workman begins with the seed, and that great care is taken to procure seed of the very best quality. "In the spring of the second year, when the seedlings are about eight inches in height, they are staked with bamboo-canes and tied with rice straw, the plants being bent in different desirable shapes. In the next fall they are transplanted to a richer soil, and are well fertilized. In the following spring the plants are restaked, and twisted and tied in fanciful forms. This mode of treatment is given until the seventh year, when the trees will have assumed fairly large proportions, the branches being trained in graceful forms, and the foliage like small clouds of dense green. The plants are now taken up and placed in pots one and a half feet in diameter, and are kept well watered every succeeding year; great care must be taken to keep new shoots pinched back. After another three years of this

JAPANESE MAPLE.

treatment the trees are virtually dwarfed, there being no visible growth thereafter. Maples form one of the best materials for the artistic fancies of the Japanese graftsman. Many times a great many different varieties are grafted on one stem. Seedling maples are spliced and tied together when growing. After they have formed a union the desired shoot is cut off—this is kept up until ten or twenty varieties are obtained. Maples thus grafted form lovely features for lawns, their varying hues and types of foliage enhancing each other's beauty.

"The æsthetic idea shows itself in every line of Japanese industry, and especially is it the case with our nursery and landscape gardeners. The most inexperienced need not fear any difficulty in our mode of gardening if he but uses his mind and efforts in the right direction. The skilful artist introduces into his miniature garden, not regular geometrical forms, but anything odd, irregular, and artistic. To us gardening is not mathematics, but an art; hills, dales, rivulets, waterfalls, bridges, etc., vie with each other in presenting their quaintest forms and fancies and harmonious symmetries. Dwarfed plants of all descriptions deck the scene here and there in thousands of peculiarly artistic shapes. We derive lessons from Nature, and strive to imitate her as much as is practicable, although on a smaller scale."

The Japanese maples belong to the natural order *Sapindaceæ*, and, though differing so widely from most of their congeners in general appearance, are genuine members of the great maple family so widely known throughout the world. And yet they occupy a field in

Acer—Japanese Maples.

horticulture peculiar to themselves, being both curious and beautiful beyond comparison. Though real trees, they are known to us only in shrub-like proportions, and, being of slow growth, are adapted to use in localities where other sorts would soon become too large for the situation. It is true that they do not carry conspicuous blossoms, but the foliage is resplendent in crimson, pink, yellow, red, purple, and gold, each in varying shades, so that when rightly planted in masses they become as attractive as any bed of flowers, with the advantage of maintaining their charms throughout the entire summer. Their habit of growth is low and somewhat bushy, ranging from three to fifteen feet in height, with well-formed and shapely heads. With our present information it is impossible to class them as to species with exactness, and, though this has been often attempted, scarcely two authorities are entirely agreed. This, however, is a matter of small importance in practical horticulture, as all the general features are well understood and appreciated. The following are the best sorts, and probably the only ones at present before the American public, to be commended for cultivation :

A. japonicum.—This is a species growing ten to fifteen feet in height, with bright green leaves. These are larger than in most of the sorts represented in this group, nearly round, and scalloped or fluted, though not deeply indented. The flowers appear in early spring, are delicate pink and decidedly attractive. This is all the more noticeable as in most of these maples the blossoms are comparatively inconspicuous. The tree is of slow growth, holding its foliage well into autumn and growing more and more beautiful with age. It has given off a variety, *A. j. aureum*, whose distinguishing feature is in its yellow or golden leaves, which are intermingled with varying shades of green, productive of the best effects, the colors being retained through the entire season. The tree are hardy as far north as southern New England and western New York. Another variety, less striking, is *A. j. aconitifolium*, aconite-leaved, the foliage of which is deeply cut, giving it a very picturesque appearance. There are also a large-leaved sort, *A. j. macranthum*, one with small foliage, *A. j. microphyllum*, and still another, *A. j. scolopendifolium*, smaller than the last, with wavy or scalloped leaves.

A. polymorphum.—This is supposed to be the parent of many of the remarkable varieties that have come to the knowledge of the public, and that now enter so largely into horticulture. Whether it be true, as some assert, or otherwise, that it was the original form, it has certainly given off a number of varieties which in garden and lawn planting have no superiors, and, indeed, few or no equals. It rises from twelve to eighteen feet, as observed

ACER POLYMORPHUM.

in this country, and can be grown in tree or bush form as may be desired. The foliage is rather small, deeply lobed, and has a habit of taking on different forms as well as colors, the young growth materially differing from the older on the same tree. In the fall special tints are assumed, which add much to its attractiveness. The bark is smooth, and, all things considered, it is one of the prettiest small trees that can be planted in ordinary locations.

A. p. atropurpureum.—This is known as the dark purple-leaved Japan maple, and is probably more largely planted than any other variety. It is identical with the preceding except in the color of the foliage, though perhaps not of quite so vigorous growth. Planted in groups, few things are more effective, and it retains its color well into autumn. Where contrasts are desired, nothing serves a better purpose and harmonizes so well with the other plants of its class. *A. p. atropurpureum nigrum* has still darker leaves.

A. p. atropurpureum dissectum.—This is a dwarf, pendulous form of the most exquisite beauty, and, though not found in most gardens, is worthy of a place in all. The bark on the smaller branches is crimson, and the foliage is so finely cut as to give it the appearance of beautiful lace or hanging draperies. The leaves are bright rose-color when young, but as the season advances change to a dark purple, retaining, however, their beauty to the last. At its best the tree grows to a height of ten to twelve feet, but as usually seen in American gardens is scarcely more than five or six. Mr. H. H. Berger, the California florist and importer, who saw the plant in its native habitat, says: "In the English Park at Yokohama, on the Bluff, is to be seen one of the most perfect specimens of this kind. The stem is about twelve feet high, and is completely masked by the drooping branches. When the wind sweeps through them, the sunlight playing with the purplish and red tints of the leaves, they are like a cascade of rich, royal lace, or a poem in color. This sort has a sub-variety, the crimson, fern-like leaves, variegated with pink, green, and white, being extremely delicate and of a dwarf growth." *A. p. albo variegatum* is another form with marked peculiarities. It has small, deeply cut leaves variegated with white and yellow, but is not esteemed one of the best of the family. This and *A. p. cristatum* are seldom seen in the gardens, though both are pretty plants and worthy of more general use.

A. p. rosco-pictum.—In this the leaves are equally finely cut, but variegated with white, yellow, rose, and green, making it in springtime a wonder to all beholders. The shrub, though so delicate in appearance, is reasonably hardy where other varieties are known to thrive. It is still scarce in the markets, so that the demand for it has never been fully met. *A. p. rosco-marginatum* is of slow growth, with small leaves curiously tipped and edged with red or rosy-pink. There is a variety of this differing only in the more intense colors of the variegation. *A. p. versicolor* is another form in which the foliage much resembles that of the parent *polymorphum*, but is spotted with pink, white, and green. It is a strong grower and one of the most interesting sorts.

A. p. reticulatum is very much of a dwarf. It is

Acer—Japanese Maples. 55

described as having deeply lobed leaves, traversed over a whitish ground with a network of translucent, yellowish-green lines, and as coloring finely in autumn when it is at its best. It contrasts well with the *A. p. sanguineum*, or blood-leaved variety, which in turn is one of the pret-

CUT-LEAVED JAPAN MAPLE.

tiest of the whole family, and one of the most largely planted. This is at its best in June, and, though it later loses some of its brilliancy, it is a gem of rare worth to the end of the season. No massing or grouping is complete without it. It is also especially effective as a single specimen, and is one of the most popular of all the family.

Ornamental Shrubs.

SHEPHERDIA—Buffalo Berry.

THE shepherdias constitute a small group of the Oleaster family, which was named by Nuttall in honor of John Shepherd, at the time curator of the Liverpool Botanic Garden. All are natives of North America, having a range from New Mexico to the British Possessions on the north, but mostly confined to the western sections of the United States. But one, only, proves of especial value as ornamental or possessed of economic worth. *Shepherdia argentea*, popularly known as the buffalo or rabbit berry, grows from five to fifteen feet high and is of slender proportions, with opposite ovate or oblong leaves silvery white on both sides. The branches are covered with gray bark, the whole contrasting pleasantly with surrounding trees or shrubs of darker shade. It is slightly thorny and capable of being used for hedges to advantage. The flowers are small and in compact clusters, bright yellow, appearing in April or May in great profusion. The fruit which follows is still more beautiful. It is scarlet or crimson, and hangs in bunches sometimes covering almost the entire bush or tree, and contrasting with the light-colored foliage to great advantage. These berries have an agreeable acid taste and prove edible for man or beast. The plant was introduced many years ago to English gardens where it was received with favor becoming its merits. *S. canadensis* is a slightly smaller shrub but far less valuable for garden cultivation. Its fruit is less showy, quite insipid, and the branches, young leaves, and indeed the whole plant are more or less covered with rusty scales. But it has its

Ligustrum—Privet.

uses, as it grows on gravelly banks and in sterile soils where little else will thrive. Its hardiness is proverbial, as it holds its own against the most adverse influences.

LIGUSTRUM—Privet.

THE privets are among the most useful as well as most ornamental of the small trees or shrubs known to cultivation. They belong to the olive family, *Oleaceæ*, and comprise from twenty-five to thirty species with numerous varieties of especial value. Some are natives of Europe, others of temperate and tropical Asia, and still others of Australia and the South Sea Islands. Those most familiar to our northern gardens are nearly all deciduous, but farther south the evergreens predominate. In either case, the bright green foliage and somewhat conspicuous white blossoms and berry-like fruit show to advantage. Few plants are adapted to a wider climatic range or will thrive under such diverse conditions. All the species are of easy growth and comparatively indifferent to soils and situations, though as they are rapid growers they should be supplied with plenty of vegetable food for best effects. It is claimed that they are especially indifferent to the smoky atmosphere of large towns and cities where so many others fail, and this, taken in connection with the fact that they are notably free from insect pests and from disease, suggests their more liberal planting in such locations.

L. vulgare is the common privet or prim of the old English gardens, and is still largely in use for hedges and fencing. It is a native of Europe, including the British

Islands, and so is adapted to the climate and well fitted to the purposes to which it is applied. It has smooth, elliptic-lanceolate foliage, and, when properly treated with that end in view makes a compact, well rounded bush somewhat in tree form, from six to ten feet high. The blossoms are white and in compound racemes, coming out in early summer, and are both numerous and fragrant. These are followed by nearly black berry-like fruit in clusters of no particular beauty. In the hedge-rows, where the plants are kept well cut in, but few flowers are seen as they are borne on the new wood. But when grown on the lawn or in the border they are among the most floriferous of shrubs, the odor of the blossoms when in close contact being so intense as to be offensive to many persons, a characteristic of nearly all the privets. There are several well established varieties, some of which may be preferred to the typical form. Of these *L. v. buxifolium* has leaves resembling those of the common box; *L. v. frutulutcum* has yellow fruit and more dense foliage; *L. v. variegatum*, leaves margined with yellow in one form, and marked with white in another, and others have more or less noticeable variegations.

L. ovalifolium is popularly known as the California privet, though for no good reason, as it is a native of Japan and not of the State whose name it bears. It is preferred to the *vulgare* because of its larger foliage and more rapid and vigorous growth. The leaves are oval-elliptic or obovate, and much more persistent. Even in the vicinity of Boston they often keep their color until midwinter, while farther south they are practically ever-

Ligustrum—Privet.

green. Unfortunately this species is not quite as hardy as some of the other forms, and so cannot be depended upon in northern New England or the Northwest. In Newport, Rhode Island, it is the favorite shrub for hedges and is extensively planted along the seaside, and often in most exposed situations, where it seldom suffers from wind or cold. It appears to be peculiarly fitted for growth near the shore, and is much used for the protection of other and less hardy forms. It is found, also, on the open grounds and in the borders of the most pretentious villas of that city of palaces. There is a variety catalogued as *L. o. tricolor* whose foliage is beautifully marked with green, yellow, and white, the combination continuing the entire season with all the distinctness of the first growth. There is, however, a tendency on the part of some of the more vigorous branches to turn to the original color, and it sometimes becomes necessary to cut out the green-leaved branches in order to preserve the full beauty of the bush. It is often grafted on privet stock several feet high, and grown as a ball or a pyramid, when the effect is very fine.

L. japonicum.—This is an evergreen species from Japan, reaching a height of six to ten feet, and is a vigorous grower with oblong-ovate foliage more sharply pointed than in most privets. The leaves are thick and glossy and are capable of withstanding northern winters better than most of the others. The variety known as *macrophyllum* has still larger foliage, but is scarcely to be preferred. Neither the type nor its varieties are much known in the United States and cannot confidently be recommended except for the southern or middle sections.

L. ibota is another species from Japan, and is in all respects one of the best. It proves a somewhat smaller form with even more slender twigs and branches. The flowers are in drooping racemes in midsummer, and so fragrant that they perfume the air for a considerable distance. It is able to endure a greater degree of cold than the last mentioned, and so is to be preferred in northern latitudes where the ordinary privet hedge is considered precarious. It is believed to be as hardy as the old English privet which Mr. W. C. Egan of Highland Park, Chicago, puts down as the only ligustrum which thrives in that locality.

L. lucidum.—This also belongs to the evergreen branch of the group and is not freely planted in this country. The leaves are much larger than in either of the others described and are more oval. Its flowers appear in wide-spreading panicles in early autumn, and are quite showy as well as fragrant. It is a native of China, though found growing freely in Japan also, rising to a height of ten to twelve feet.

HYPERICUM—St. John's-wort.

THE hypericums constitute a large genus of tender and hardy herbaceous plants, shrubs, and trees. The Greek name appears to have been originally applied to a species growing freely in Egypt and southern Europe, which was but twelve to eighteen inches high. The best of the hardy sorts growing as shrubs or small trees are natives of North America, though Europe furnishes several that are valuable. They can be easily

Hypericum—St. John's-wort.

grown, and are worthy of more attention than has been accorded them in promiscuous cultivation. Thus far, of the many members of the family but few have come into common use as ornamental, but as their merits are better understood they are fast growing into popular favor. This is seen in the fact that, whereas a few years ago the hypericums were seldom advertised by the nurserymen, they have now so far won upon the public as to secure a place in almost every catalogue. Nearly all have opposite leaves, which are frequently dotted with darker colors, giving them a somewhat singular though not an especially attractive appearance. The sap or juice is usually acrid and disagreeable to the taste. There are now before the public a dozen or more species or well-defined varieties, from which selections may be made that are sure to please.

HYPERICUM KALMIANUM.

H. kalmianum, popularly known as St. John's-wort, takes its name from the Swedish botanist, Peter Kalm, who first saw the plant on the borders of Niagara River, during a visit to this country. It was introduced to European gardens about the middle of the last century, and was received with much favor. In England, especially, it has long been regarded as among the best of ornamentals in the department to which it is particularly adapted. It is a low, spreading bush from two to four feet high, the diameter of its top being often greater than its height. The leaves are numerous and somewhat crowded, linear, slightly glaucous, and about two inches long. The blossoms appear in August. They are glossy yellow, somewhat resembling in their general appearance those of the dandelion, and are very numerous and long-continued, and afford a marked contrast to the deep green foliage and the prevailing colors at that time of the year. For planting singly or in masses this is very effective, and is especially adapted to small lawns or gardens. It also has the merit of doing better than most plants when growing in partial shade.

H. aureum is of even less size than the preceding, and in some respects is to be preferred to it where a small shrub is called for. It is a recent introduction, and a worthy candidate for popular favor. Though diminutive, it produces much larger flowers than the *kalmianum*, and in equal profusion. Of the few hardy shrubs blossoming in early autumn this is certainly one of the most showy. The flowers appear in August, and continue their brilliancy, under favorable conditions, until October.

H. prolificum is a larger shrub than either of the preceding, growing four to six feet, and with a wide-spreading

HYPERICUM MOSERIANUM.

head. The branches are covered with light red bark, which separates easily into a multitude of thin scales. The flowers are large and showy, produced in terminal clusters, and continuing from July to September.

H. densiflorum is a shrub four to six feet high, with numerous slender branches and a multitude of small but conspicuous yellow blossoms in compound cymes. It is comparatively new to cultivation, but bids fair to win a place among our most popular sorts. It adapts itself readily to sandy and comparatively barren soils, where it often thrives; but, like other plants, prefers more favorable conditions.

H. adpressum is almost a creeper, and grows wild in southern New England, preferring moist locations. As it is indigenous to Nantucket and the islands of Vineyard Sound, it will undoubtedly thrive anywhere at the seaside where a low plant is desired for covering bare spots or for edgings. It seldom rises above two feet. The leaves are oblong, acute, and thin; the flowers, bright yellow, and covering the bush during most of July and August.

H. patulatum is also a spreading St. John's-wort. It is a native of Japan, and is probably, all things considered, preferable to the preceding, though having the same general characteristics. The flowers are bright yellow, opening in the summer and continuing until early frosts.

H. buckleyi is a native of the mountainous regions of the Carolinas, and, though hardy in most parts of the United States, is but little known. It is described as a small shrub with slender branches covered with loose, reddish bark. The leaves, green above and paler beneath,

Hypericum—St. John's-wort.

are two inches or more in length, oblong, rounded at the apex and narrowing to the base. The flowers, terminal and solitary, an inch in diameter, are very pretty.

H. moserianum.—This is acknowledged to be one of the very best low-growing plants of recent introduction. Its habit is free and graceful; it produces long, slender, much-branched stems, leafy to the base and drooping towards the ends. It is free-flowering, each blossom measuring from two to two and a half inches in diameter, in color a rich golden yellow, which is rendered still more effective by the numerous yellow stamens and crimson anthers. It blooms continuously the entire season, and whether used as a bedding plant, for borders, or as single specimens, is equally desirable. There is some question as to its hardiness as far north as New England, where it sometimes fails to withstand the winters unless afforded protection. But in the Middle and Southern States it is sure to prove of the highest value. Its height is from two to three feet, and it is of very easy cultivation.

H. oblongifolium.—This is sometimes catalogued as *H. hookerianum*, which may be regarded as a synonym, as the two certainly represent the same plant. The flowers are large and bright yellow, though the bush is not as free a bloomer as some of its class. They are sufficient, however, to make it very desirable in the shrubbery or border, where midsummer blossoms are especially desired. It grows but about two feet, and is nearly an evergreen and quite so in the Southern States. In the North it needs slight protection, and does best in half-shaded situations. It is a native of Nepaul. Others which may be mentioned as desirable are *H. elatum*, a native of the United States, which rises about three to four feet, whose blossoms are small but produced in great profusion on the numerous slender-growing branches, and *H. hircinum*, which grows to about the same height and has paler yellow flowers, with very long stamens whose prominence gives the bush a somewhat peculiar appearance and makes it quite attractive. It is a product of the Mediterranean countries.

PHILLYREA.

A GENUS of *Oleaceæ* consisting of but four species. These are all natives of the Mediterranean countries, and have long been known but not freely employed in cultivation, thus helping swell the large list of neglected plants that are worthy of a more generous recognition. Two kinds only are reputed sufficiently hardy for general cultivation in the United States.

P. media is a vigorous-growing bush of a spreading habit, ten to twelve feet high, with lanceolate leaves, entire, veiny, and of a very distinct appearance. The flowers are white, appearing in early spring and followed by fruit in one- or two-seeded drupes of long continuance. It is not suited to the New England climate, but in the southern Middle States would prove an acquisition. There is a pendulous variety of much merit, sometimes known as *olæfolia*, or *ligustrifolia*, which is worthy of notice.

P. vilmoriniana is counted more hardy and is coming into favor in English gardens. Its leaves very much resemble those of the Portugal laurel, being large and leathery, dark green tinted with bronze on their first appearance. It produces large, fragrant, white blossoms in axillary clusters, in May or June. It does well in any light friable soil and can safely be planted in our Southern and Southwestern States.

ERICA—CALLUNA—Heath.

THE ericas are said to include more than four hundred species, some say nine hundred, and these have in turn given off almost numberless varieties. Most of these are too tender for out-of-door cultivation even in the Southern States, and only a very few are found to thrive in more northern localities. They are largely natives of the southern hemisphere, especially of Africa, and but very few are adapted to temperate climates on either side of the equator. Under favorable climatic influences they retain their foliage during the winter and so are mostly classed as evergreen, but in this country they scarcely answer to that description, though even here their foliage is persistent, usually holding on until late autumn. The flowers are mostly nodding, axillary, or terminal, produced in fascicles and in many colors. The typical form of the blossom is tubular with the mouth somewhat contracted, and from one to four inches in length.

As a rule, the ericas require more care and attention than most other plants in ordinary garden cultivation, and are more particular as to situations. In all cases a peaty soil is preferred, though any good friable compost such as will answer for the rhododendron or the azalea will serve a good purpose. A soil of stiff clay or one impregnated

Ornamental Shrubs.

with lime proves fatal to success. But with all these drawbacks it is surprising that the heaths are not more generally planted in this country than they are. Equal and even more attention is bestowed on many sorts in no wise superior, and with less satisfactory results. Their flowering season covers almost the entire year, though no single variety is a perpetual bloomer. But by a proper selection from among the various forms these little plants become objects of interest and pleasure in spring and summer, autumn and winter.

ERICA FRAGRANS.

As greenhouse and conservatory plants many of them are highly esteemed, and whole houses are often devoted to their cultivation. Both in the greenhouse and garden there is wide range for selection. Some of the species are small and especially suited to growing in borders and edgings. Others are larger and well adapted to massing or appearing as single specimens. Some blossom in early spring, others in midsummer to late autumn and even in early winter. The numbers of species and varieties are so great that no attempt will be made at description except in a general way. Nor will any attempt be made at scientific accuracy in names and qualities.

Almost every one has heard of the interesting little shrub known as the Scotch heath or heather, and though, botanically, this is classed as calluna it is such an impor-

Erica—Calluna—Heath. 69

tant member of the family that the popular conception may well be recognized in this connection. *C. vulgaris*, the Scotch heath, grows from one to three feet, with purplish flowers, disposed in long, terminal, spicate racemes, from

HARDY ERICA.

July to September. It is common to nearly all northern and central Europe, especially on the hillsides and in waste places which, by its presence, are made beautiful. Unlike many of the more delicate ericas, these are able to endure many hardships. They grow in thin soils and thrive in exposures that few plants are able to endure. Some of the varieties are especially beautiful and should be selected for garden cultivation in preference to the original form, whenever they can be procured. There is a white-flowering plant that is very pretty; a flesh-colored, one much admired; and others with double blossoms, though these are not always improvements. There are also varieties with golden and silver-colored shoots, that are exceedingly attractive both in flower and foliage, and especially adapted to planting in masses, and the wonder is that they are not more freely used. In short, all the heaths known as callunas are worthy of attention.

RHAMNUS—Buckthorn.

THE buckthorn family, *Rhamnaceæ*, is widely distributed throughout Europe, northern Africa, Asia, and America, and, though preferring a warm climate, most of the species thrive also in the temperate zones. It is believed that the lotus fruit spoken of by Homer was a product of a member of this family, though not that which is now known to us in cultivation. The buckthorn proper, rhamnus, constitutes a tribe of more than thirty species, a few only of which are found in the United States. They are all large shrubs or small trees with opposite leaves, minute, fragrant flowers, and stony fruit.

The species known as the common buckthorn, *R. catharticus*, is found growing wild in New England, though Emerson expresses the opinion that it was probably

Stephanandra Flexuosa.

brought here from Europe and has made its way from the gardens into the fields and woods. It grows from ten to twenty feet high, with a smooth stem and gray or olive-tinted branches. These are numerous, stiff, and well supplied with thorns. The leaves are ovate, notched, and marked with hairy veins beneath, though smooth on the upper surface. The flowers are small but numerous, growing in clusters, of various colors but mostly pink and white, and are followed by berries which become black when ripe, and hang late into the autumn. The buckthorn is much planted for hedges in England, and, when properly trained, constitutes an almost impassable barrier to man and beast. It makes a fairly good ornamental plant, though not in general use for that purpose, partly, perhaps, because of its slow growth.

STEPHANANDRA FLEXUOSA.

THIS is a genus of *Rosaceæ* of but a single species. It is closely related to the spiræas and very much resembles some of the most valuable members of that family. It is supposed to be of Japanese origin, and is certainly an interesting low-growing shrub, seldom more than six feet in height. The small branches are numerous and the foliage dense and compact. This is distinguished by being deeply and finely cut or toothed, and taking on a purplish-red tint in its young growth and again in autumn, being of a rich glossy green during the summer. The flowers are small, but so numerous as to cover the whole bush late in June or the early part of July. Many regard it as one of the best of recent introductions, as it serves a good purpose whether planted singly, in groups, or in front of taller growths. Such plants, beautiful in both foliage and flower, are to be preferred to many of those which possess but a single virtue.

COLUTEA.

THE coluteas constitute an interesting genus, of the order *Leguminosæ*, which have long been known to Old-World cultivation and are now somewhat widely distributed in America. They are all rapid growers and of easy cultivation, being not over particular as to soils and situations, though thriving better on dry land than in peat and excessively wet locations. In some European countries they were at one time much planted as ornamental hedges, though now superseded in that respect by newer and better adaptations. The genus is not large, but it exhibits peculiarities of such a striking character that specimens may well be planted in every considerable garden.

C. arborescens.—This is doubtless the best known and most widely distributed member of the family, and everywhere answers a good purpose in decorative planting. In some countries it is known as the bladder senna, from medicinal uses to which the leaves have been put. It is a native of Italy, where it grows to a height of six to ten feet and forms a rather open but well-rounded head covered with glaucous green foliage. The leaves are compound, with seven to nine elliptic leaflets, which hold their

Colutea.

somewhat peculiar green color well into autumn. The flowers resemble the famous sweet pea in form, but are less conspicuous. The blossoms are yellow, and continue in succession during the entire summer. They are succeeded by bladdery pods, two to three inches in length, containing the seed, which also hang long, so that we have the bush in flower and fruit at the same time and for a protracted period. These yellow blossoms and the thin, almost pellucid pods, hanging among the green leaves make it an object of interest to almost every one who comes into its presence. This bush is one of the few which brave the terrors of Mount Vesuvius, growing to the very summit, and is found occasionally even within the circle of the crater, where vegetation can scarcely gain a foothold. Such a fact ought to suggest its adaptation to dry and sandy plains and other locations which it is difficult to cover with herbage. No one would be likely to suspect that such a slender-looking shrub would withstand such exposures and thrive where so very little else can endure. There is a variety of much smaller dimensions, known as *C. a. pygmæa*, which is also a shapely bush and may be of service where space is limited.

C. cruenta is much the same as the preceding, except that it grows only from four to six feet and has blossoms tinted with pale red or blood-color. These also appear in early summer, and are continuous in succession for a long time. The leaflets, which are from seven to nine in number, are smaller and more glaucous that in the preceding. It is a very pretty plant. *C. hallepica* is another form with larger yellow flowers than either of the other species. It grows from three to five feet. *C. media*, as its name indicates, is intermediate in its proportions, and has bright orange-yellow flowers. There are few or no varieties in cultivation sufficiently distinct to call for special mention.

CRATÆGUS—Thorn.

IT has been said that the thorns produce a greater variety of beautiful small trees and shrubs than any other family. This may be true, but it will not readily command universal assent. Still the numerous species and varieties possess peculiar attractions and some of them have long been noted for their excellences. They appear both as shrubs and trees, and can be easily trained and shaped to suit the purposes of the propagator. Some of the best known species are natives of North America, and have been carried to Europe and widely distributed. Others have been brought from the Old World to the New, and have been received with equal favor. Nearly all the species are beautiful in leaf, flower, and fruit. The botanical name, *cratægus*, originally given to the hawthorn, is derived from the Greek *kratos*, signifying strength, and is fitly applied.

English hawthorn, *C. oxyacantha*, has been famous in England for many generations and is also well known throughout all northern Europe. Grown in upright form, it makes a well-shaped tree, sometimes fifteen to twenty feet in height. It bears the shears well, and can be kept within such small proportions as may be desired, and shaped at will. It is distinguished for its rigid stems,

Cratægus—Thorn.

numerous sharp spines, and attractive foliage. The leaves are obovate, deeply lobed and toothed, somewhat wedge-shaped at the base, smooth, and glossy. The flowers are white, very fragrant, and mostly arranged in corymbs. These are followed by deep red fruit which hangs long on the bush and is quite showy. In Europe the hawthorn is extensively used for hedges, not only for its beauty but because it furnishes protection alike against man and beast; but it has never been popular as a hedge plant in America. The species takes its popular name from the berries which in England are called haws. There are several varieties of much value, the best of which are as follows: *C. o. rosea*, pink-flowering, differs from the type chiefly in the color of its blossoms. In this case the petals are rose-colored, or pink, with the tips or claws bordered with white, and a well-grown tree in full bloom never fails to secure admiration. *C. o. rosea superba* has larger petals, which are dark red or crimson without the white tips, and is very fine. There are several double-flowering forms, one of which, *C. o. flore pleno albo*, has large white, double flowers in great abundance, shading to pink before they fall. *C. o. coccinea* duplicates the last by producing similar blossoms

FLOWERING BRANCH
OF CRATÆGUS OXYACANTHA.

in pink or scarlet, and a variation from this, known as Paul's double-scarlet, marks still another advance. In each of these the individual flowers are miniature rosettes and are very interesting. What is known as the Glastonbury thorn, *C. o. præcox*, is so named because of its supposed origin at Glastonbury Abbey. It is remarkable from the fact that it flowers much earlier than the original. In England its blossoms often appear at Christmas and again late in the following summer. In the United States it is one of the earliest to put forth its foliage, appearing at its best farther south than New England or New York. Numerous other departures from the original appear, some with different-colored fruit, and others having foliage variegated, sometimes with yellow and sometimes with white; but none with characteristics so peculiar in these respects as to call for special notice.

The American hawthorns, though usually spoken of simply as thorns, are also rich in species and varieties. They include about one third of all the kinds known, and are almost invariably hardy and worthy of cultivation, though such as are natives of the Gulf States should not be transplanted to the North where the winters are severe. Among the best of these are the following: *C. crus-galli*, cockspur thorn, is pronounced by some authorities the best of all American sorts, but there is little reason for giving it such especial prominence. As a small tree, it grows freely in almost any good soil and is worthy of general cultivation. The stem is erect, throwing off branches in whorls, which grow almost at right angles with the trunk. This gives the head a stratified appearance and adds to its

HAWTHORN ON OUTSKIRTS OF A WOOD

attractions. The leaves are dark green, almost glossy above, thick and inversely wedge-shaped or obovate. In autumn they assume a rich coloring. The flowers are white tinged with pink or red. Fruit, round, edible, and of a scarlet or sometimes a dull red color, hanging on the branches far into the winter. The species will do better in partial shade than most other sorts. It grows from ten to twenty feet, and blossoms in June.

C. coccinea, scarlet thorn, is a native species of great value, spreading over a wide range of territory and everywhere much admired. It is a small tree rising from ten to twenty feet, and in every way well proportioned. The numerous wedge-shaped, thin leaves are bright, soft and pleasant to the eye. The bark on the stem is rough, with a grayish cast, but on the smaller branches it is often smooth and of an olive-green or reddish shade. The spines are strong and sharp, well calculated to resist intruders. The flowers are white and pretty, appearing in early June, and followed by bright scarlet berries that hang long and present a striking appearance. It is their prominence which gave the name to the species by which it is properly known.

The species has given off a large number of varieties, some of which prove of horticultural worth. This is especially true in the case of *C. c. macrantha*, which by most authorities is ascribed to this origin, though some are disposed to raise it to the dignity of an independent species. Professor Bailey points out several qualities in which it diverges from the *coccinea*, among which is the fragrance of its blossoms. He says that in this respect it is

Cratægus—Thorn.

entirely different from the scarlet thorn, which possesses anything but an agreeable odor. "A bush of the long-spurred thorn when in flower scents the air for a considerable distance." Other reasons are given which appear to be convincing, but they need not be discussed in this connection. This long-spurred thorn is found growing in a state of nature from the banks of the St. Lawrence to Minnesota, and proves perfectly hardy throughout all the Northern and Middle States; and it will probably adapt itself to even more southern situations.

What is known as the white thorn is supposed to be also a variety of the scarlet, and by others it, too, is raised to the dignity of a species. It is known as *C. mollis* and is one of the largest and most conspicuous of the tribe. *Garden and Forest* says: "In cultivation the white thorn is a beautiful plant, of rapid growth and good habit, conspicuous in winter for the whiteness of its branches and for the number of its large chestnut-brown shining spines. The flowers, with the exception of those of one species of the Southern States, are the largest produced by any member of the genus. The leaves are large and of a lively green, and the fruit, which is as large as that of a small crab-apple, is brilliant scarlet with a conspicuous bloom."

C. pyracantha fructo luteo.—This is an evergreen thorn, and one of our most desirable plants for the lawn or garden. As it is not large, it is well to plant three or five so that they will combine in appearance as one plant, when they will make a well-rounded, bushy form, ten feet, or perhaps more, in height, and as many in diameter as may be desired by the planter. The leaves change to a brownish purple in the winter, and are very effective in such situations. In fact, there is scarcely any shrub that makes a better appearance on the lawn, taking the year as a whole, than this. As a hedge-plant it is even more especially desirable. It is true that its growth is slow and it takes some time to secure a good stand, but when the result is obtained nothing in the line of hedges is more beautiful. The branches are small, numerous, compact, and run into each other to such an extent that the whole line appears as though it might be a single growth. In spring and summer, the glossy foliage is covered with fragrant white flowers, followed by bright orange or scarlet berries which add much to its attractiveness. *C. p. lalandii* is a variety with larger leaves, and even more beautiful, but, unfortunately, like the type in the North, it is not entirely hardy. The evergreen thorns cannot be relied upon beyond the limits of southern New England. *C. pyracantha cuneata* is cultivated under the name of

CRATÆGUS SPATHULATA.

Magnolia.

cratægus spathulata. It has bright scarlet fruit which hangs on during the winter among the purplish, persistent leaves, making it a very ornamental plant.

MAGNOLIA.

THE magnolias constitute an interesting family, and as a whole are unsurpassed among hardy trees in beauty of foliage and flower. They are widely distributed both in the Old World and the New, and in higher and lower latitudes, some being deciduous and others evergreen. As classified by the botanists there are about twenty species and an equal or greater number of varieties known in cultivation. Some are natives of tropical Asia, and others of the colder portions of the temperate zone. A few are found in the West Indies, New Zealand, and Australia, and a much larger number come from China, Japan, the Crimean and Himalayan Mountains. Several are natives of North America and indigenous to the United States. But, wherever found, their attractions have been such that the better sorts have become common property throughout the horticultural world.

A large number of species and varieties blossom in early spring before the foliage appears. In this case the flower-buds are formed and fully grown the previous summer, and so are ready to respond to the first genial breath of spring. It is astonishing how quickly these trees are clothed with flowers as the winter wears away, and how fully the promise of the last year's growth is fulfilled. None of the American species are of this class, as with them, together with many foreign varieties, the foliage precedes the flowers, though all are early spring bloomers. Only a few are evergreens, but the leaves of many of the deciduous species are so persistent and hold on so long in autumn that it might not require a very great change in climatic conditions to enable them also to be clothed with living green. In most cases the leaves are large and inclined to be fleshy, alternate, entire, and sometimes dotted with pellucid spots, giving them a rich and attractive appearance. The fruit usually grows in a cone or hard, compact cluster, which becomes scarlet or bright red during the period of ripening. As the seeds drop out they may often be seen hanging by a slender thread several inches below the bunch in which they matured. They consist of small, hard nuts covered with a pale-red, fleshy substance which should be removed before planting. Without observing any prescribed order or classification, the following list is given as including the more desirable hardy forms adapted to cultivation:

MAGNOLIA GLAUCA.

M. glauca, known in some localities as sweet bay, and in others as swamp laurel, is the only species that is known

Magnolia.

to be indigenous to New England, its northern limit being in the swamps of Cape Ann near the sea. It grows freely in the southern Middle States, where it appears as a well-formed, small tree, and can be safely transplanted as far north as Canada. The foliage is good, the leaves being elliptic in form, from three to five inches in length, dark green above and whitish beneath. The flowers are white, composed of nine delicate petals tapering at the base, and arranged in three circles which unfold in succession. They appear in May or June, are fragrant and of long continuance on the branches. Though not one of the most conspicuous members of the family it is worthy of more attention in garden planting than is usually accorded to it.

M. conspicua, known also as the *yulan*, as it was introduced from a Chinese province by that name, is one of the most beautiful of all the hardy magnolias, and has come to be planted accordingly. It grows to be a tree in somewhat shrubby form, from twenty to thirty feet high, and with numerous branches. The foliage is good, the flowers large, pearly white, and produced in such profusion as to almost hide the stems and branches from observation. They are cup-shaped, from seven to nine inches long and three to five across, appearing just as the leaf-buds begin to open. I have counted twelve hundred of these great blossoms upon a single tree growing in Newport, which for many years has not once failed to be such an object of beauty as to attract visitors, who never weary of admiring and praising it. The tree is reasonably hardy, but does best when planted on the southerly side of a wall, where

84 Ornamental Shrubs.

the wood is better ripened and the large autumn buds more fully developed.

MAGNOLIA CONSPICUA.

M. hypoleuca, sometimes called the silver-leaved magnolia, is a native of Japan, and is one of the most striking and attractive members of the family. According to Professor Sargent, who studied it in its native forests on the island of Yezo, it sometimes rises to the height of more than a hundred feet, and is a valuable timber tree, though in garden cultivation it does not appear to have anywhere

Magnolia.

reached such proportions. It is emphatically a northern species, and may not be at its best in locations where winters are not more or less severe, and where the ground is not covered with snow a portion of the year. This will certainly recommend it to a large constituency. The leaves are from twelve to twenty inches long, and seven or eight inches broad, bright green on the upper surface and pale steel-blue or silvery white on the lower. The flowers are from five to seven inches in diameter, with creamy-white petals and brilliant scarlet filaments. They are very fragrant, and appear after the foliage is fully expanded.

M. kobus is also Japanese, common to the forests in that country, and of but recent introduction to garden cultivation. In its native habitat it grows to the height of from seventy to eighty feet, with a straight trunk nearly two feet in diameter and covered with slightly colored bark. The head is described as pyramidal in old specimens, round, and with short, slender branches. The flowers appear early and in advance of the foliage. They are white, slightly tinged with yellow, from four to five inches across, and without special fragrance. The leaves are obovate, bluish green, six or seven inches in length and about half as broad. It has the bad reputation of not blossoming when young like some others, so that in planting for early effects one should procure as old and large trees as possible. It appears to be hardy, and even in the absence of flowers is a desirable acquisition.

M. salicifolia.—In his visit to the Japanese forests, Professor Sargent found, and describes, another magnolia, to which the above name has been given. He speaks of it as growing on mountains two to three thousand feet above the level of the sea, and also as found in swamps, as well as in dry situations. It thus appears on Mount Hakkodo, where it is a slender tree fifteen to twenty feet high, with ovate, acute leaves, light green above and silvery white below. These are some six inches long, two inches broad, and borne on slender, short petioles. As he did not see the tree in flower—and it is not known to have blossomed in this country or Europe—the peculiarities of the flower are not fully known.

M. soulangeana is one of the hardiest members of the family. It is an importation from China, and supposed by some authorities to be a hybrid, produced, through natural causes, between *M. conspicua* and *M. purpurea*. The tree does not grow so tall as the former, but forms a low, spreading head, and produces immense blossoms, white, with purple at the base, affording a very attractive combination. It has the advantage of blossoming later than the *conspicua*, and so helps maintain a succession. *M. speciosa* is also a Chinese hybrid. The flowers are somewhat smaller than those of the *soulangeana*, appear still later, and remain longer on the tree. They are red and white, or rose-colored, and afford a marked contrast with some of the others described. This variation from the prevailing colors and the period of blossoming makes it especially desirable in grouping.

M. stellata.—Few small trees or shrubs are more beautiful than this, whether planted singly or in groups of three or four. It is a low-growing species from Japan, seldom rising more than eight feet, with spreading branches which

Magnolia. 87

in early spring are crowded with white blossoms in advance of the foliage. The leaves are from three to five inches long, elliptic in form, and abundant. The flowers are also small, some three inches across, with about fifteen narrow petals, slightly reflexed, encompassing a cluster of bright yellow stamens. They are somewhat star-shaped, fragrant, and of longer continuance than those of most of the species and varieties. The plant is of slow growth, but has the advantage of coming into blossom when very small, and, under favoring conditions, every season. It is claimed to be the earliest bloomer of all the magnolias, as well as the most profuse. Its usefulness in the garden is sometimes impaired by late frosts and heavy rains, as it does not always wait for settled weather. But this is true of nearly all early flowering plants, such as insist upon crowding the season. No one will make a mistake in planting this magnolia, however small his grounds. Specimens less than two feet high often produce flowers freely, and so apparently out of season when contrasted with the surroundings as to be of especial interest. The plant is hardy, and thrives in ordinary soils, preferring, however, leaf mould and peaty substance with plenty of moisture. This magnolia has long been a favorite with the Japanese flower lovers.

M. watsonii, newly introduced to the public, comes from Japan, and is a well-formed tree, producing obovate leaves five to seven inches long and about three inches wide. These are bright green above, veined and margined with yellow. The under surface is a paler shade of green, and, especially in the younger growth, covered with silky hairs. The flowers are white, five or six inches in diameter, and highly fragrant. They are especially beautiful because of the blood-red filaments which surround the pistils, as well as from their large proportions. They grow singly on short peduncles, and cover the tree while the foliage

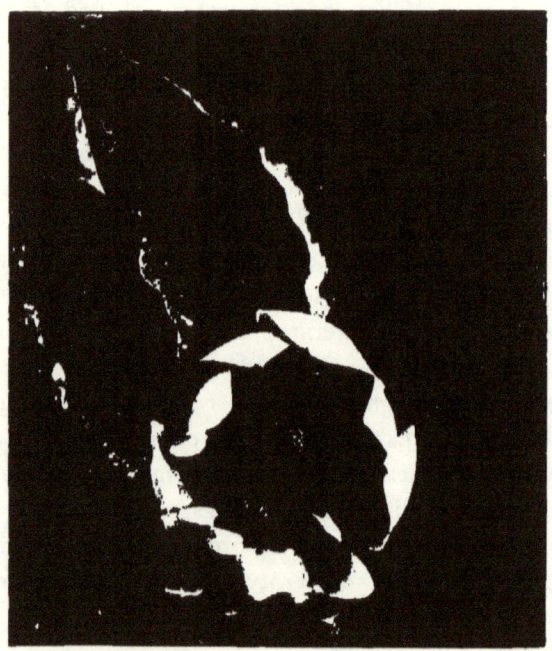

MAGNOLIA WATSONII OR PARVIFLORA.

buds are breaking into green. *M. lennei* is another striking Chinese hybrid, whose flowers are crimson or purple without, but white within. It is prized wherever known, and claimed by some to be the best of the purple varieties.

M. purpurea, popularly known as the purple magnolia,

because of the color of its flowers, is also of Chinese extraction. Downing says that both the white and the purple varieties "eclipse every other floral object, whether tree or shrub, that the garden contains." This variety is said to be a hybrid, and one of the more hardy sorts. The blossoms are white within and dark purple or lilac-colored without, and very fragrant. In all other particulars it partakes of the general characteristics of the family. *M. atropurpurea*, or dark purple magnolia, blooms in May and is distinguished by having the darkest-colored flowers of the whole list. It, too, is from China, and not at all common in the nurseries or in garden cultivation. It is to the magnolias what the purple beech is to its congeners.

M. macrophylla, or the great-leaved magnolia, has immense leaves from one to three feet long and ten to twelve inches wide. The flowers are nearly or quite a foot across, with white petals, purple at the base, and moderately fragrant. The tree is medium sized, and has the most tropical appearance of any of the hardy forms. The combination is such that it is difficult to decide whether the leaves or flowers are most to be admired. It is an American species, and sufficiently hardy for planting in favorable conditions in the Northern and Eastern States.

M. alexandrina is small, rising from ten to twelve feet, and is more of a bush than a tree. It is one of the earlier-blooming varieties, bearing large pink flowers in May, and in some favored localities late in April. It is more especially adapted to gardens and grounds of limited dimensions. *M. gracilis* is also shrub-like in its proportions, being not more than ten to fifteen feet high, and producing purple flowers of deeper hue than most others of its class. They expand about the first of April or in early May, and are especially valuable for massing or use in borders of shrubbery. There is no doubt as to its hardiness in ordinary situations and its adaptability to various garden soils. *M. norbertiana* is another of the shrubby kind, though somewhat taller, with good foliage and pink blossoms appearing two or three weeks later than those of the last mentioned. Though good in itself, it is not especially to be preferred to the other pink varieties. All these low-growing sorts should be more generally cultivated than they are now, as they may well take the place of plants which, though better known, are far less valuable where early effects are desired.

M. obovata.—This is a rather tender species from Japan, and, though growing but five or six feet high and in bushy form, produces large and showy blossoms with six petals and very sweet-scented. The leaves are large, obovate, deep green, and of good substance throughout the entire summer. It is not counted entirely hardy, even in the Middle States.

M. grandiflora—Evergreen Magnolia.—This is, perhaps, the most beautiful and interesting of all the magnolias, but, unfortunately, it is not sufficiently hardy to withstand the rigors of our northern winters. It is indigenous to nearly all the Southern States and in some instances becomes a tree from seventy to eighty feet in height, though its average proportions in cultivation are much less. In all cases, it is inclined to an erect, slender, and somewhat pyramidal form, and thus is a most desirable

Magnolia.

tree for street planting, and the wonder is that it is not more largely employed in beautifying both streets and parks wherever it can be available. The leaves are large, oval-oblong, coriaceous, and bright glossy green on the upper surface, though somewhat rusty beneath. The blossoms are also large, six to eight inches in diameter, white, with from nine to twelve petals. They are deliciously fragrant, and continue a long time in perfection. There has recently been more or less discussion as to the northern limits where this magnificent tree may be grown. It used to be said that it could not be depended upon much north of Richmond, but it has been demonstrated that fine specimens can be maintained not only in Washington but even in Philadelphia. Mr. Thomas Meehan gives an account of a tree about thirty feet high, in one of the parks in the latter city, which blossoms and ripens seed every year; and adds that "it is not the only one, by any means, to be found in the city. In Fairmount Park, near Horticultural Hall, there is a tree which has been flowering and seeding for many years. I could name a dozen trees thriving hereabout, several of them of flowering age. I would without hesitation lift the line of its hardiness from Richmond, Virginia, and set it at Philadelphia." Good specimens are reported even farther north, as at Dorosis on Long Island, but they have to be attended to with especial care, such as only experts and enthusiasts have patience to apply. Mr. Meehan reports having seen them growing in England north of London, but in almost every instance they were trained on the side of a building, flat and fan shaped. There are a number of varieties of this tree, but very few, if any of them, are to be preferred to the original form.

ABELIA.

THE abelias constitute a small genus of very ornamental shrubs of the order *caprifoliaceæ*, and, though not sufficiently hardy to withstand the winters of the northern parts of the United States, they are especially adapted to garden cultivation throughout the southern half of the Union. They will probably thrive in most locations south of Washington, through the Southwest and California, and with but little or no more care than is afforded many other choice plants that do not surpass them in interest or beauty. All are of easy cultivation and showy in both leaf and flower. The name was given to the genus in honor of Dr. Clark Abel, a noted physician and botanist, who was attached to the British embassy to China in 1817, and who probably first directed the attention of English horticulturists to their merit. But it does not appear that the plants were brought to England until Mr. Robert Fortune entered so largely upon his work of introducing the botanical treasures of the flowery kingdom to the European gardens.

A. rupestris.—This species was brought from China to England by Mr. Fortune in 1844, but, from its real or supposed inability to withstand the climate of that country, was largely treated as a conservatory or greenhouse plant. It has more recently been found equal to the demands in some of the southern counties, and where, according to a writer in *The Garden*, of London, it is

Abelia.

seldom injured by cold or frost. In this country it is found to thrive below the latitude of Washington with as little care as is given to many other plants in common use. It is known to do well in Philadelphia, and good specimens are occasionally seen in the vicinity of New York, facts which fairly indicate its northern territorial range. At its best it makes a well-shaped bush from five to eight feet high, and usually about as many in diameter. When desired it can be grown on a pillar or trained to a wall as a low climber with good effect. The plant is clothed with small, oblong, glossy foliage, and in its season is abundant with blossoms. The tubular flowers are of a pale rose-color without, and white within, continuing from July or August, according to location, until frost. *A. r. grandiflora* is a variety originated in Italy, which is said to be a decided improvement upon the type. The flowers are much larger and more beautiful, the color remaining the same, and the whole plant is more robust.

A. serrata.—This is another of the broad-leaved evergreen species from China, and is of about the same proportions as the last. It is, however, an early spring bloomer, producing its terminal flowers singly, but in sufficient abundance to cover the entire bush. They are very large, pale red, and exceedingly fragrant. It has the reputation of not being quite so hardy as the *rupestris*, but as being especially desirable in southern localities and for greenhouse cultivation.

A. triflora is a native of Hindoostan, and grows to a height of five or six feet. Its foliage also remains through the winter months, like that of the rhododendron and the kalmia, and the plant should be subjected to much the same conditions in cultivation. The flowers are terminal, in threes, yellow tinged with pink, the sepals being long and clothed with hairs. It flowers in autumn and is one of the best of the late bloomers. *A. floribunda* is of American origin, supposed to be a native of Mexico and adjacent territory. It is a smaller plant, growing about three or four feet, with opposite long leaves and damask or rosy-purple blossoms nearly two inches long, in clusters at the ends of the numerous twigs. These are very showy and appear in early spring.

A. spathulata.—Though of later introduction this species is coming to be as well known and as fully appreciated as most of its predecessors. It comes from Japan and proves to be a much-branched and free-flowering evergreen shrub that is certain to attract attention wherever grown. The leaves are from one to two inches long, elliptic-lanceolate, slightly toothed, glabrous above, somewhat rough and hairy beneath, the edges being marked with purple. The flowers are nearly or quite sessile, in pairs, the corolla being an inch long, white within and marked with blotches of light yellow on the throat, appearing early in spring. These are so numerous as to envelop the whole bush, and are of long continuance. It is believed to be adapted to more northern localities than most of the other sorts, but as it is not yet widely tested, its hardiness under adverse conditions is not fully determined.

Rosa Rugosa.

ROSA RUGOSA.

THE roses are not often classed among shrubs, but this is justly entitled to that distinction. Though a true rose its shrubby characteristics could not be well overlooked in this connection. And when properly

ROSA RUGOSA.

grown and cared for, it will be found one of the most valuable as well as most ornamental plants in the whole list. It grows from five to ten feet and, if permitted, will attain a diameter nearly or quite equal to its height, having a tendency to send up new stems from year to year so that there is scarcely a limit of possibilities in this direction. At the same time it is easily kept within desired bounds, while the fresh growths may be transplanted with entire success. The foliage is abundant, dark green, plicated, and dense through the entire summer and autumn. Were it not a flowering plant at all, it would still be desirable on the lawn or in the border. The blossoms are single, from three to four inches across, deep red with yellow stamens, showing abundantly in June and July, and more sparingly in midsummer and autumn. These are followed by large heps of scarlet-crimson which continue long into the autumn, and in the opinion of many are as beautiful as the blossoms themselves. They form quickly, after the flowers, in succession, have dropped their petals, and so it is common to see both fruits and flowers in profusion at the same time, and, as both are terminal, the combination is especially effective. The bush is too strong a grower to be suitable for the ordinary rose garden, its proper place being on the lawn or in the border. In the gardens at Newport, *Rosa rugosa* is more freely planted than any other shrub, and particularly in exposed situations. It is seen in many villas on the highest cliffs, where it bravely withstands the fiercest winds that come in from open sea. In such situations the plants are often cut down sharply on the approach of winter, as they should always be when the foliage is desired near the ground. Left to themselves they become coarse and bare at the base, while if reduced in height the plant retains its

ROSA RUGOSA HEDGE AT THE SEASIDE.

beauty at all times. The rapidity of its growth is such that if cut to one or two feet, it is sure to answer all the purposes of a comparatively low ornamental shrub the following season.

There are several varieties and numerous hybrids worthy of notice. *R. r. alba* is much the same except in the color of the blossom, which is pure white and very attractive. The plant is slightly less vigorous and of smaller proportions, but the scarlet heps contrast so well with the flowers and the rich green foliage as to make it especially desirable. *Madame Georges Bruant* is a hybrid with broad and handsome foliage and pure white flowers in clusters, semi-double and quite fragrant. It blooms at intervals throughout the entire summer. *Agnes Emily Carmen* is a cross with Harrison's yellow, and is one of the best of the group. The blossoms are deep crimson, semi-double, borne in clusters and appearing at intervals during the entire season, even to the coming of frost. The foliage is also good. The plant may not be quite as vigorous and as well adapted to rough exposures as the *rugosa* proper but it may be put down as reasonably hardy throughout the Northern States.

There are other roses that may be grown in bush form, and are especially adapted to that treatment, but they do not call for description in this connection.

MORUS—Mulberry.

THE mulberries belong to the bread-fruit tree family, *Atrocarpc*, which includes some of the most interesting of nature's products. Though more especially adapted to the tropics, some of the species appear freely in the temperate zones, and are almost as highly prized by civilized races as others are by the savages who gather their daily food from the stems and boughs within their reach. The tribe includes, along with the poisonous upas, the famous banyan tree of India, the celebrated cow tree of South America, the fig tree, and many others especially adapted to the wants of man and beast. The mulberry appears in many countries, and some of the forms are indigenous to eastern Asia, southern Europe, the United States, and Canada. The several species are curious and interesting, and nearly all of them are of especial interest to the botanist and practical horticulturist. They have been in cultivation from the earliest antiquity, and are mentioned in the Old Testament Scriptures as objects of interest and almost veneration. Many of them serve a good purpose in furnishing food, and as ornamental plants; and such might be cultivated to advantage much more generally than they now are. The hardy species are easily grown and long-lived. They produce sweet and juicy fruit, though this is not equally palatable to all people. Mulberries were first introduced into England in the year 1548, and afterwards became so popular that "the mulberry gardens" were a prominent feature of some of the best estates. These plantations were util-

Ornamental Shrubs.

ized in furnishing food for silkworms, as well as producing fruits for the table, and for the manufacture of

RUSSIAN WEEPING MULBERRY.

wine. All the species are late in putting forth their leaves in spring. The foliage, when it does appear, is a bright,

Morus—Mulberry.

dark green, and contrasts finely with the colors of most other trees in the vicinity.

The red mulberry, *M. rubra*, is the only species indigenous to New England. It is a medium-sized tree with large, rough, heart-shaped leaves, sometimes serrate and sometimes lobed. The flowers are of a greenish-yellow tint, small and numerous, followed by dark red fruit, sweet to the taste and preferable to that of most other sorts. The wood is hard, strong, and very durable, and is often used in boat building, and for posts whenever it can be obtained in sufficient quantities. Though not strictly a first-class tree for ornamental planting or for purposes of forestry, it is worthy of a place in every large collection.

The black mulberry, *M. nigra*, was carried from Persia to Europe in the 16th century, and thence brought to America, where it has been cultivated to a limited extent in gardens and private grounds. The foliage is much the same as that of the preceding, and the fruit, in the form of a spike composed of numerous calyces and carpels, is succulent, and, to many people, of pleasant taste, but not to all. It is said to be a very long-lived tree and to grow well in most parts of the country. The white-fruited mulberry, *M. alba*, is a well-known silkworm species, at one time very popular, but now much neglected. It is a medium-sized tree with succulent leaves growing in great abundance. It is worth growing only as a curiosity. The Spanish mulberry, *M. hispanica*, has large, smooth leaves, and from the vigor of its growth and its rich purple fruit is by many preferred to all others for garden planting.

Ornamental Shrubs.

PYRUS JAPONICA—Japan Quince.

THIS popular shrub, formerly known as *Cydonia japonica*, was brought to England as early as 1815, where it was received with much favor and thence distributed on the continent. Wherever it is known, it is recognized as one of the best of the many good plants that the Island Empire has yet given to the western world. Still its merits do not appear to be fully appreciated by many, not to say most, of the planters of the present day. Nicholson in his *Dictionary of Gardening*, quotes approvingly an earlier authority to the effect that it is "one of the most desirable deciduous shrubs in cultivation, whether as a bush or on the open lawn, trained against a wall, or treated as an ornamental hedge plant. It has also been trained as a standard, and in this character its pendant branches and numerous flowers give it a rich and striking appearance, especially in spring. It is difficult to unite with its congeners by grafting; but if it could be grafted high on the pear, the hawthorn, or even the quince, it would form a most delightful little tree. It is readily propagated by layers or suckers, and also grows by cuttings."

There are numerous varieties growing from five to eight feet in height, and if trained with that end in view, nearly or quite as many in diameter. Some of these are scarcely known to the general public but all are worthy of consideration. Probably the most perfect collection is to be found at the Arnold Arboretum (Harvard University) and they are described in brief by C. J. Dawson, the superintendent, as follows :

Pyrus Japonica—Japan Quince.

"The type, *Pyrus japonica*, has a very dark red flower of considerable size; the foliage takes on a purplish tinge and the habit is upright. *P. j. moorlosi* is an extremely fine variety. In habit it is low with arching branches, in fact, almost pendulous, the foliage very slender and narrow, while the medium-sized flowers are pink and white in color. *P. j. wallardi* is of medium good upright habit with flowers of the very darkest crimson in color. *P. j. atrosanguinea* is similar to *moorlosi*, only the habit is not so airy, the flowers are not borne so abundantly, and the leaves are much larger and not so narrow.

"*P. j. macrocarpa* is of a splendid spreading habit with dark foliage and medium-sized flowers of a light carmine-red color. *P. j. foliis rubris:* The foliage is decidedly colored, the flowers salmon-red in color, while the habit of the plant is very compact and not of the average height. *P. j. roseo flore pleno* has good semi-double deep rose flowers of large size. *P. j. versicolor:* Flowers pink and white, the habit of the plant being very compact. *P. j. atrosanguinea plena:* In this variety the habit is very dwarf and compact and it is a decidedly free bloomer. The flowers, of a deep red, are only slightly double. *P. j. grandiflora:* One of the best; flowers very large, pink and white in color, and very fine airy habit. *P. j. umbellata:* Also a very good variety, with flowers of deep rose color borne very abundantly. *P. j. nivalis* is a splendid white variety, laden with medium-sized flowers. *P. j. simplex alba:* The flowers a trifle larger than *nivalis*, the last two being both excellent varieties."

Pyrus maulei, a more recent species of Japan quince,

is also attractive; it is much dwarfer, seldom growing more than three feet in height, and very compact in habit. It is more covered with thorns than the japonica type, and the colors take on a different tone of reds and pinks than do the older Japan quinces. Its various forms make one of our most beautiful of recently introduced shrubs.

It is to be regretted that with all their good qualities these plants are not more freely used for hedges, as they are certainly superior in many respects to the privets which are now so freely employed. Their foliage takes on various shades of color as the season advances, from olive to pink, the latter appearing often in the new growth after cutting back or trimming. Under such circumstances the hedge seems crowned with scarlet, answering to the show of flowers in early springtime. Such a hedge is not so easily broken down as those composed of less thorny plants, nor will it require so much cutting as though composed of shrubs disposed to larger proportions. It may be as cheaply planted and as readily grown, the cost of keeping in order being less. Besides, the owner would have the satisfaction of breaking in upon the monotony that now threatens the almost exclusive use of a single type.

PYRUS MALUS—Flowering Apple—Crab.

THE apple blossom is always beautiful, and yet but comparatively few persons think of growing the tree simply as an ornament for field or garden. Perhaps in the minds of some the very fact that the apple is one of our most common as well as most useful fruits

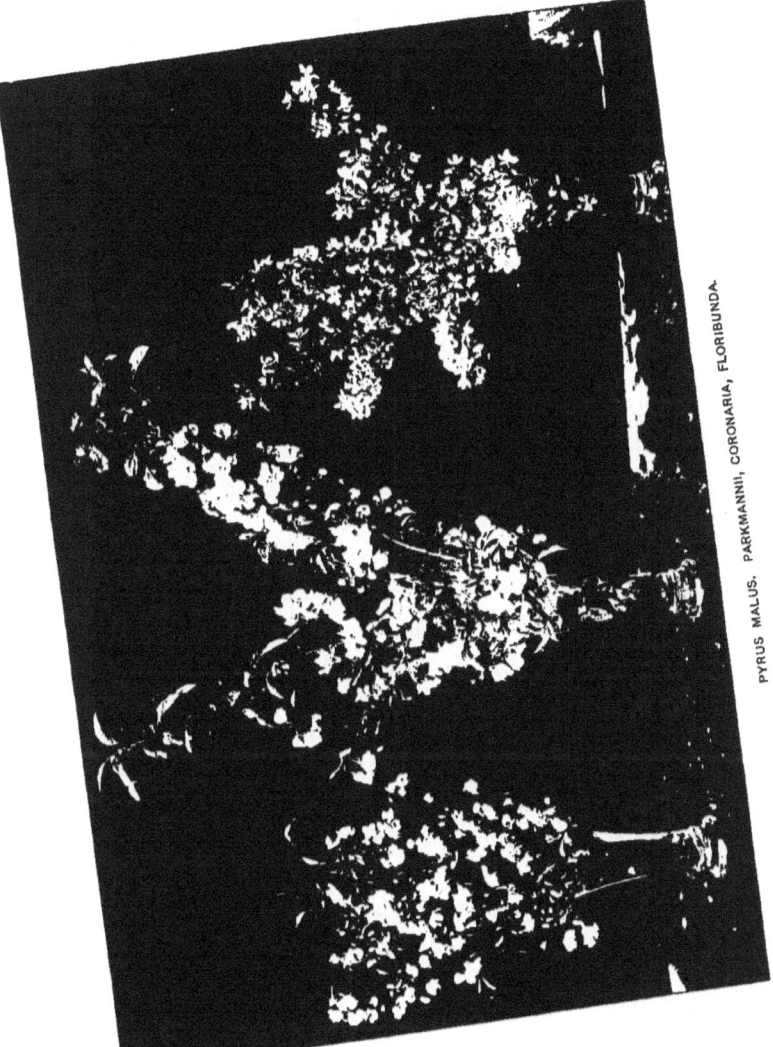

PYRUS MALUS. PARKMANNII, CORONARIA, FLORIBUNDA.

shuts it out of consideration in this connection. But there are species and varieties that are both rare and ornamental in the highest degree, and that need only to be known to make for themselves a place in every considerable collection where best results are desired. To some of the most desirable of these attention may well be called. In their wild state they are almost universally known as crabs, and as such are indigenous to most of the temperate regions of both Europe and America.

P. m. coronaria, the American crab, is a native of the United States, and outside of cultivation is probably found at its best on the Alleghany Mountains, where it appears as a small and shapely tree, growing to a height of about twenty feet. Though perfectly hardy, it has not shown itself to be very aggressive, as its range is quite limited. Still, it is found, though in comparatively small numbers, in locations much farther north. The beauty of its blossoms early attracted attention, as they are large, rose-colored, and very fragrant; the odor, in the opinion of many persons, resembling that of the common sweet violet. The fruit, though not of much economic value, is also fragrant and interesting; the apples are small and exceedingly numerous, and hang long on the branches. Some of the supposed varieties are still more beautiful than the type, especially those with variegated foliage. That known to the trade as *P. m. aucubæfolia* has leaves which are distinctly marked with white, and in some cases shaded with pink, and is very desirable. Another, known as the double white-flowering apple, *P. m. alba plena*, produces large double blossoms which are very sweet-scented.

Pyrus Malus—Flowering Apple—Crab.

This is much smaller than its parent, growing but five or six feet high.

A Chinese crab, *P. m. spectabilis*, grows to a height of twenty to thirty feet, and has large, pale red or rose-

DOUBLE FLOWERING APPLE.

colored, semi-double flowers in April or May. They are nearly sessile, and appear in umbels. The leaves are oblong, oval, and smooth, and give the tree a somewhat showy appearance at all times. The fruit is not especially good. The variety known as *flore roseo pleno* produces double rose-colored flowers nearly two inches in diameter.

These appear in May, and are also very fragrant, making the plant in every way desirable for ornamental purposes.

P. m. floribunda, the Japanese flowering apple, is one of the most interesting acquisitions that has been made to our list of ornamental trees in many years. It grows five to six feet in height, has small, obovate leaves, and produces beautiful, rich, rosy-red blossoms in great abundance in early spring, and sparingly throughout the summer. The shoots are slender, and often bend beneath the weight of the small apples which are borne on long stems, and it is difficult to tell whether the shrub is more to be desired for its appearance when in flower or in fruit. Of this *Garden and Forest* says, editorially: "This, it seems to us, is the most beautiful of its race, and one of the best ornamental plants in cultivation. It is particularly beautiful before the flowers expand, when the bright red flower-buds cover the branches. The Japanese crab should be planted in rich soil, and allowed plenty of room in which to spread its wand-like branches. * * * Improving with age, the Japanese crab grows more beautiful every year; the severest winters leave it uninjured, and insects and diseases pass it by. The variety with bright pink, semi-double flowers, known as *Pyrus parkmannii*, is equally beautiful, though it is a rather less hardy plant."

The Siberian crab, *P. m. prunifolia*, has been much planted, and has also many good qualities. The white, single flowers often cover the entire tree, and give it a showy head in April or May. The fruit when ripe is yellow, with the side toward the sun showy red. The tree is of larger growth than most of its class, rising some

Pyrus Malus—Flowering Apple—Crab. 109

twenty feet, and is suitable for either the orchard or the garden.

What is called Bechtel's crab is a new American product, and is already regarded as one of the best that

BRANCH OF FLOWERING CRAB.

has made its appearance. It is supposed to have sprung from the western form of *P. m. coronaria*, having originated at Stanton, Ill., and has been put on the market within a very few years. It produces large double pink blossoms, much resembling small roses, and in great abundance. They are also exceedingly fragrant and said to answer the purpose of cut flowers, retaining their good qualities for a long time. Unlike most of the crabs, the blossoms do not appear until the foliage is well advanced, the bush being thus in leaf and flower at the same time, a decided novelty in this class of plants. The tree is a rather slow grower when young, but has the habit of blooming when very small, often when not more than two feet high. This, in connection with the fact that it lengthens the season of apple-blossoming nearly two weeks, makes it especially desirable in connection with the other and earlier sorts.

BERBERIS—Barberry.

THIS is an interesting group of hardy plants, with an Arabic name, though indigenous to many sections of Europe, Asia, and America, as well as Arabia. A few are evergreens, but by far the greater number are deciduous, growing in bushy forms and in almost every kind of soil, though not favorably disposed to low, marshy situations. It is said that there are some fifty species, to which may be added several varieties of special value and well known in cultivation. Some are but a few inches in height, with round, compact heads, while others grow to ten or twenty feet, specimens occasionally appearing in tree form rather than as bushes or shrubs. They have, as a rule, yellowish wood and inner bark, ovate and pointed thorny foliage slightly serrate, and numerous yellow flowers. The fruit is mostly scarlet or crimson, and so intensely acid that birds will not eat it; but, properly prepared with sugar, the berries of some

Berberis—Barberry.

species make excellent preserves and syrups for the table. The roots and sometimes the bark are used for the production of a yellow dye used in coloring. Both the root and bark, as well as the leaves, are esteemed valuable for their medicinal qualities. A peculiarity of the flower is that some of its parts are possessed of a remarkable degree of irritability, so that if the filaments are touched on the inside with even the point of a needle, the stamens are thrown down upon the stigma, and the petals incline in the same direction, showing what appears to be a wreck of the entire floral structure. But the seeming ruin is not permanent. Equanimity is soon restored, and the several parts slowly resume their places, when the flower lives on as though nothing unusual had occurred.

The common barberry, *Berberis vulgaris*, a native of Europe, is usually a low, bushy shrub, but capable of being trained into almost any form desired. It produces its bright yellow flowers in May or June, and they are followed by small, oblong, acid fruit. The branches are provided with sharp spines, and the leaves are also pointed with bristles, making the shrub difficult to handle. When planted in rows and properly cut in, it makes an almost impenetrable hedge against man or beast. It is a long-lived plant, notwithstanding its diminutive size. This shrub is so widely distributed throughout the country as to lead to the supposition that it is a native of the soil. But it is not. Having been brought here and planted by our forefathers, it kept pace with the growing population, and having in a measure escaped from civilization, it planted itself along the roadsides, passed over the walls and fences, and in some instances took possession of entire fields and hillsides to the exclusion of the forms of plant life that had long had possession, somewhat after the manner of the conquests made by the pale-faces in their strifes with the natives. There is a variety that has richly colored purple foliage, and that proves very effective in planting, either by itself or in combination with other sorts. Its general characteristics are much the same as those of the type, and, whether planted singly, in masses, or in the border, it is equally good, and capable of affording most desirable contrasts. But for the best results it must have plenty of sunshine and not be grown in too moist a soil. Standing side by side with yellow-leaved plants, the combination of purple and gold is all that can be desired. Unlike many of the so-called foliage plants, it holds its color from spring to autumn, and can be used on large or small estates to advantage.

The American species, *B. canadensis*, was so named by Pursh, the distinguished botanist, but is not a native of Canada or even of New England or New York. It is indigenous to the Alleghany Mountain region, thence southward to the Gulf and to some sections west of the Mississippi. In general it is much the same as the *vulgaris*, but with botanical differences sufficient to maintain a specific classification. It is a more diminutive shrub, having smaller and less bristly, pointed leaves, fewer-petalled flowers, and less conspicuous fruit. But in the general outline to the unscientific eye the two are much the same. It is entirely hardy, and has been found able to withstand intense cold, provided it is favored with a well-drained soil. In many sections the farmers believe that the presence of either of these barberries causes rust in wheat and perhaps other growing crops, as the under side of the leaves is often of a brown or rusty color. But it is known that rust is largely a fungous growth, and that the fungus is of such a different character that the disease, even when existing in the same vicinage, could not have been transferred from one to the other. There is probably, therefore, no good reason for the widespread prejudice among the agriculturists against these plants.

The box-leaved barberry, *B. buxifolia*, is so named from the resemblance of its foliage to the common box famous in old-fashioned gardens. It is also known as the sweet-fruited barberry, *B. dulcis*, and some have supposed that the two names designated distinct species, but this is not the fact. This shrub comes from the Straits of Magellan, and is counted an evergreen, though in very cold climates it is not strictly such. The leaves are oblong, smooth, and glossy, without hairs or spines, and about half an inch in length, with very short footstalks. The cup-shaped, amber-yellow flowers appear very early, almost before the winter is past, and are borne on slender, pendulous stalks in great abundance, and followed by dark-colored fruit. Under favorable circumstances the shrub grows to the height of six to eight feet. Darwin's barberry, *B. darwinii*, grows but about two feet, and produces an abundance of orange-colored flowers in May, and sometimes again in autumn. The deep purple berries are oblong and about an inch in length, and armed

Ornamental Shrubs.

with teeth. The species has the advantage of being an evergreen, and, as the branches are numerous and the foliage dense, it is showy in winter as well as in summer.

The Japanese barberry, *B. japonica*, is a compact shrub, seldom growing more than two to four feet, and having

BERBERIS JAPONICA.

unbranched stems covered by a grayish bark. The compound leaves are about three inches long, dark green above and lighter beneath, and composed of from seven to nine leaflets. They are armed with slender but sharp spines, and not easily handled. In autumn the foliage assumes brilliant shades of orange and scarlet. The

Berberis—Barberry.

flowers are large and in terminal racemes three inches in length. This species is claimed by some to be the most beautiful of all the barberries, though not so widely known and generally cultivated as some others. It is found to be reasonably hardy in the Northern States.

Another Japanese species, *B. virescens*, was introduced in this country in 1849, and, after long testing, was offered by nurserymen in the market. The flowers are small, in short racemes, and yellow tinged with green, the fruit oblong, compressed, and purple-scarlet in color. The first specimens brought to Europe came from an elevation of nearly nine thousand feet above the level of the sea, and the species ought to be hardy in any ordinary locality. It is highly praised by those who know it best. The many-flowered barberry, *B. floribunda*, is also a native of Asia. Its yellow blossoms are in pendulous racemes, in which they hang somewhat loosely, appearing in June. They are pretty, though not especially striking. The leaves are obovate, long, and tapering toward the base almost to a point. The shrub attains a height of six to ten feet.

Thunberg's barberry, *B. thunbergii*, came from Japan, and is a most valuable acquisition. Though not yet largely distributed, it is to be found in not a few gardens, and is everywhere looked upon with especial favor. It is said that the Japanese prize it not only as the best of its family, but also as one of their most charming plants. The bush grows five feet, and has numerous slender stems and branches, some of which are upright and others almost horizontal or even pendulous, all being

Ornamental Shrubs.

armed with small but stiff and sharp spines. The spoon-shaped leaves are small, dark, and glossy, green in spring and summer, and in autumn take on a variety of hues—crimson, orange, and bronze—which are retained for some weeks. The flowers are solitary, distributed along the branches, and of a lighter shade of yellow than those of most others. They appear in early spring,

FRUITING BRANCH, BERBERIS.

and are followed by an abundance of fruit which hangs from the under side of the branches from one end to the other, and covers the whole bush. The oblong berries are bright scarlet in color, exceedingly showy, and never fail to give the shrub a most charming appearance in late autumn and early winter. It is a peculiarity of these berries that they contain very little pulp or juice, and so do not shrivel or even wrinkle after they ripen, even though subjected to frost and repeated freezing. They are very persistent, and retain their places far into, and sometimes through, the winter, and up to the time of blossoming for the next season. This alone would cause the plant to be most highly esteemed, as it is not often that a shrub is found which is almost equally attractive in spring and summer, autumn and winter. Whether standing as a solitary shrub on the lawn or in the garden, planted in groups or placed in the border or hedgerow, *B. thunbergii* never fails, when thus fruited, to brighten

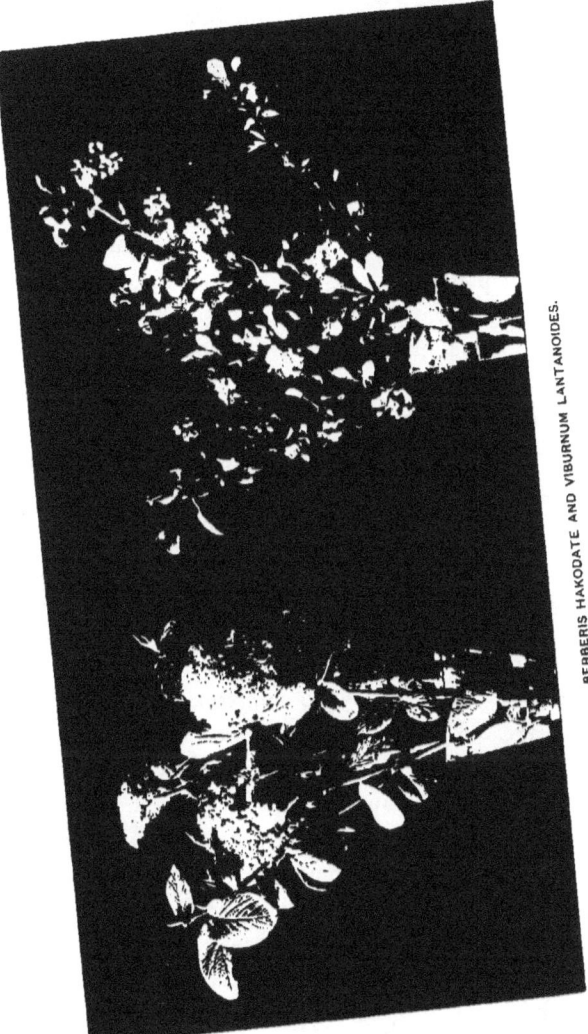
BERBERIS HAKODATE AND VIBURNUM LANTANOIDES.

the landscape and relieve the dullest months of the year of much of their monotony and gloom. It is thoroughly hardy, a good grower, and needs but little care, since it naturally assumes a good form and retains it from year to year. It may safely be put down, all things considered, as one of the very best of the barberries for ornamental planting.

An evergreen species, *B. wallichiana*, is a very showy little shrub, and quite distinct. It forms a dense, compact bush, fully clothed with large, oblong-shaped leaves of a deep glossy-green tint. The bright, clear yellow flowers are borne in May or June, and are followed by purple berries. It is highly ornamental throughout the entire year. The plant comes from the Himalayas, and appears to be entirely hardy in this country. It is sometimes known as *B. hookeri*.

Another, *B. concinna*, is a recent introduction and also a native of the Himalayas. It is a low-growing bush with bright red bark on the numerous small branches. The foliage is dark glossy-green above and lighter beneath, the leaves being very small and interesting. It produces deep yellow flowers and bright scarlet fruit, and is apparently a decided acquisition. *B. cretica*, from Asia Minor, has dense, handsome foliage, dark green, with pale yellow flowers in drooping racemes. A Siberian species, *B. emarginata*, is a small plant of upright habit, having leaves finely serrated and becoming brilliant red in autumn, making it one of the choice varieties. *B. hakodate* is a new species from Japan, a more vigorous grower than most of the other new sorts, having larger leaves, and foliage which is brilliant red or scarlet. *B. sinensis* has its fruit in racemes, and it is large and brilliant red, hanging on late into the winter. *B. illicifolia* comes from Terra del Fuego, and has foliage resembling that of the holly. It is nearly evergreen in the North, and wholly so in the South, where it proves very effective as a garden plant.

GARDENIA—Cape Jessamine.

THE gardenias are all natives of warm climates, being indigenous to tropical Asia and southern Africa, especially to the region about the Cape of Good Hope. They are delightful plants in cultivation, but are not suitable for northern gardens. They belong to the order *Rubiaceæ*, and constitute a genus of about sixty species, all evergreens, growing in shrubby form, with good foliage and large white blossoms. These last are somewhat funnel-shaped, having tubes much longer than the calyx, and being deliciously fragrant. They are especially prized for cutting, and bear the operation well. The blossoms come forward freely in succession. Whether grown in the open ground or under cover, they are of the easiest possible cultivation. The species and varieties best known in the South are all natives of China and Japan, though the list might well be enlarged and enriched by additions from other countries if desired.

G. florida is probably best known in American gardens and, with its varieties, most fully appreciated. The double white flowers are solitary, almost sessile, usually terminal, and deliciously fragrant. They appear in midsummer,

Ornamental Shrubs.

and continue in succession for a long time. One of the varieties, *G. fortunei*, is in some respects to be preferred, as it blooms somewhat earlier, with equal profusion, and is of a brighter and glossier shade of green with opposite leaves in whorls.

GARDENIA FLORIDA.

G. nitida is a native of Sierra Leone, and has white solitary flowers appearing later in October and November. The tube is narrow, seven-parted, and reflexed. The foliage is oblong-lanceolate, glossy, and attractive at all seasons of the year. It grows as a compact bush from two to three feet high. The flowers of *G. radicans* are salver-shaped, but in most other respects like others of its class. They come forward among the first of the gardenias, appearing in June. There are also several variegated varieties in which the leaves are striped and spotted and very pretty.

SPIRÆA.

THE spiræas belong to the rose family, *Rosaceæ*, and are among our best-known and most popular shrubs. The genus includes about fifty species, with numerous well-marked varieties which are perpetuated in cultivation, and some of which are greatly superior to the originals. They are indigenous to Europe, Asia, and

America, but are seldom found in tropical climates or south of the equator. The species and varieties are too numerous to be fully described in this connection, or even named. They nearly all have alternate leaves, simple or pinnate, and small white or rose-colored blossoms. These last appear in cymes, corymbs, and panicles, the parts of the flowers being mostly in fives. As hardy shrubs they thrive in almost any good soil, and can be grown with little care. Some of the species are mere herbs, dying down to the root in winter and reappearing in early spring, and others are large and vigorous-growing shrubs, assuming at times almost a tree form.

S. opulifolia.—This is one of the most prominent of the American species, the familiar "ninebark" of our swamps and lowlands. It grows six to seven feet, with a rugged stem, and loose gray bark easily peeling off, whence comes its popular name. The branches are recurved, the leaves three-lobed and doubly serrate, and the flowers white, succeeded by bladdery pods turning to purple as autumn approaches. It is altogether a good plant, and will thrive in dry soils as well as in those which are wet. What is known as the golden spiræa, *S. o. aurea*, a variety of the *opulifolia*, has bright yellow leaves, and is especially desirable as a foliage plant. No one who has seen good specimens in masses or interspersed among other sorts, will hesitate to pronounce this one of the best ornamental shrubs we have in cultivation. It grows to a similar height with the parent, and is especially bright in spring while the leaves are young and fresh. The flowers are double, appearing in June.

Ornamental Shrubs.

S. salicifolia, the meadow-sweet, is one of the best-known of the smaller American sorts. It grows freely

SPIRÆA TOMENTOSA.

in moist places, on the borders of meadows or the edges of swamps, and is a low shrub of from two to four feet.

Spiræa.

The leaves are oblong and glabrous, two to three inches in length, with serrated edges. The flowers are in upright terminal panicles or cymes, and appear in July, continuing through August. It is probably more largely cultivated in Europe than in the country of its nativity. The steeple-bush or hardhack, *S. tomentosa*, is another common sort, two to three feet high, which grows freely in most parts of the United States. The stems are brown, smooth, and thickly studded with oblong leaves bright green above and whitish beneath. The flowers are in a dense, tapering panicle, spiral in form, appearing in July and continuing several weeks. They are usually of a purplish-rose color, and quite showy.

S. prunifolia flore pleno, the double-flowering, plum-leaved spiræa, is a shrub of the highest value. It was introduced to European cultivation by Dr. Siebold, who found it growing in Japanese gardens, though its native country is said to be China or Corea. It grows from six to ten feet high, in bushy form, and with numerous long, slender branches covered with smooth bark often dividing into thin scales. The leaves are lanceolate, small, and numerous, smooth above and downy on the under side, and take on beautiful autumn tints. The double white flowers come forth in early spring, covering the whole length of the arching branches. This species is more widely distributed than most of the others. *S. trilobata* is a native of the Altaian Alps, and has three-lobed foliage. It is of somewhat diminutive proportions, about two feet high, and produces a multitude of small white flowers in compact, umbel-like corymbs, appearing in May. When a small, early-flowering plant is desired, this is quite certain to give satisfaction.

S. callosa was introduced to this country from China by Mr. Fortune, and is one of the best. It is of low growth with numerous slender branches, and produces an abundance of pink or rosy blossoms in flat corymbs in June, which continue to appear through most of the summer. The variety known as the *S. callosa alba*, sometimes called Fortune's dwarf, is especially valuable. It usually forms a well-rounded head of many branches, crowded with white flowers that hang long and are followed by conspicuous bunches of seeds continuing late into autumn. For borders to garden walks or for low, ornamental hedges it is scarcely excelled. As a single plant or in masses on the lawn it is equally desirable. Another variety is the *S. callosa superba*, also of dwarfish habit and possessed of the same general characteristics, but producing greenish-white flowers in August and September. The variety known to the nurserymen as *S. callosa semperflorens* is much the same, but with red flowers instead of white.

S. thunbergii is a low, bushy shrub from the mountains of Japan, and one of the very best of the genus. It grows three to five feet, with a dense, bushy head and numerous small leaves which in late summer and early autumn take on most beautiful shades of gold, bronze, and green. Few shrubs so enliven the border or are so attractive as single specimens. No one who plants even a small place should fail to make use of this choice species. It is a very early bloomer, the flowers being in threes, not large, but so numerous as to cover the whole bush, with its

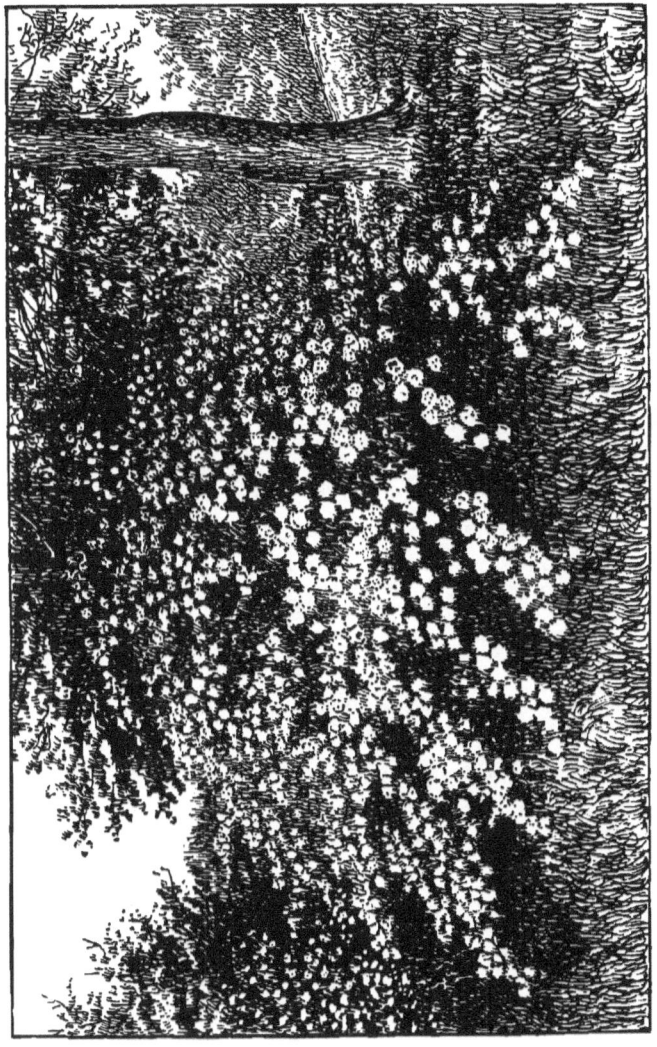

SPIRÆA TRILOBATA.

beautiful foliage as its chief attraction in late summer and autumn. It is hardy, and adapts itself to almost any soil and situation. *S. media* is taller, often growing to a height of six feet. Its greenish-white blossoms appear about the same time, and are almost equally showy, but not of quite so long continuance. It is, however, a good plant, and would be very desirable if the others were not in competition. And this will apply to the *S. hypericifolia*, or St. Peter's-wort, a kind scarcely needed under present conditions in making up a good collection.

S. fortunei has better foliage than some of the common sorts, and produces flat cymes of rose-colored or pink flowers in June. Though good in itself it is not superior to most of the others here named.

S. van houttei is a later introduction from Japan, and has been much praised. It grows in the form of a low, spreading bush with curved branches, and is from four to six feet high. While not surpassing some others as a foliage plant, its smooth, trifoliate leaves, and well-rounded form give it a fine appearance, and it is doubtful if any shrub of its dimensions under cultivation will produce a greater abundance of blossoms in the same period of time. They are white, appearing in May, literally covering every branch from end to end, so as to present much the appearance of a huge bouquet. When done flowering, the stems are almost as fully covered with the growing and ripening fruit. Nothing is of easier culture. I have taken up, with but ordinary care, large specimens when in full bloom, and replanted them without the slightest ill effect at the time or during the following season.

Spiræa.   127

S. bumalda is a very choice species of dwarf but vigorous habit. It grows two to three feet in height, with

SPIRÆA ANTHONY WATERER.

numerous slender branches. The foliage is dense and good throughout all the summer months, and when the bush is crowned with a profusion of crimson or rosy-pink

Ornamental Shrubs.

flowers it is an object of rare beauty, whether seen singly, in groups, or in the border. Few plants answer better for bedding out, as it is more showy than the geranium or the coleus, and does not need renewing every year. The blossoms appear about midsummer and continue until cold weather. *S. anthony waterer* comes to us from England as a recent production, and is presented as a variety of *S. bumalda*. It is dwarf in habit, compact, strong in growth, and perfectly hardy. The flowers are in larger heads than in the original, appearing in June, and if the old ones are removed as they begin to decay, they will be succeeded by new ones, though more sparingly, until frost. In color they are bright crimson or deep pink, and as they envelop the bush the plant becomes one of the most conspicuous objects in the garden. It grows from two to three feet, and is well adapted to edgings of borders or paths, but is never more beautiful than when planted in masses.

S. reevesii, as known in the catalogues, is a Chinese species, and one of the most beautiful-flowering sorts of the whole family. The blossoms are larger than in most of the early bloomers, of the purest white, and exceedingly abundant. But it cannot be depended upon in our far northern climate without especial care and protection, though south of New York it does well. The flowers come in round clusters early in June. There is a double-flowering variety of this species which is of much value where climatic conditions are favorable. But, unfortunately, it is probably even less hardy than the type. How far south it may thrive is scarcely yet

SUBTROPICAL GARDENING AT THE NORTH.—A VIEW AT EGANDALE, HIGHLAND PARK, CHICAGO.

determined, but it bids fair to be a boon to the gardens of that section.

S. gigantea, as its name suggests, is probably the largest member of the family. When grown in good, moist soil it reaches an altitude of eight or ten feet, with a well-rounded, bushy form. Its flowers are in large white clusters, and very effective. It is especially suited to planting by running streams and bodies of water, or in positions where a showy plant is wanted to hide obnoxious objects. Except in very large grounds it is not adapted to the border. The plant is seldom found in cultivation, or even named in nurserymen's catalogues.

S. ariæfolia is a native of the Pacific slope, ranging in its habitat from southern California to Manitoba. Though found as far north as the 49th degree of latitude, and on the Rocky Mountain slopes, it has been accounted somewhat tender in New England, and as needing slight protection in winter. This estimate of its weakness is not borne out by experiments in Newport, though it is doubtless well to give it as favorable conditions as practicable. As seen there it grows as a small shrub, with numerous branches covered with ashy-gray bark which later assumes a darker hue. The foliage is so plentiful that in a well-grown specimen the stems are scarcely visible. The flowers are individually small, white tinted with green and yellow, and in quite large terminal panicles, continuing about three weeks from the first of July. They have a peculiar odor which has been compared to that of chestnut blossoms or sweet birch. The plant is especially valuable as an under-shrub, and it grows well in shady situations, even if it does not prefer them. This characteristic makes it a valuable acquisition, for almost every owner of an estate finds places which it is difficult to cover for want of sufficient sunshine. Under favorable conditions the bush is said to grow six to eight feet high, but it is usually much lower. Some of the botanists have been inclined to consider *S. ariæfolia* as a variety of *S. discolor*, instead of being a distinct species, and some are even doubting whether it is a true spiræa at all. These are calling it *Holodiscus discolor*. A rose is just as sweet by any other name.

S. regeliana, though not widely distributed, is a good plant. It has dense panicles of pink blossoms about the first of July, and one of its distinctive peculiarities is that during the summer new stems shoot up which blossom late in the season, thus prolonging the period of flowering to autumn. It grows from three to five feet high and is hardy.

S. cana is one of the smaller spiræas, seldom rising more than two feet, and broadening its diameter into a well-rounded bush as large across as it is high. It is a native of the Croatian Alps, and appropriately named, inasmuch as the foliage takes on a grayish hue and in some cases becomes almost white. For this reason the tiny blossoms are less conspicuous than they might otherwise be, as they also are white and scattered along the stems and branches in great profusion. Where indigenous it grows freely among the rocks and in dry and barren places which there abound, and is doubtless the best adapted to such situations of all the spiræas. It is in use in England, especially in planting rockeries, for which it serves an excellent purpose and is highly prized. For some reason it is scarcely known in American gardens, but should no longer be overlooked, especially for rockwork.

S. lindleyana differs from most of the species in having pinnate foliage, with from nine to twenty-one leaflets, nearly or quite sessile, ovate-lanceolate and sharply serrated. It comes from the Himalayas, and blossoms in September, the flowers being white, very large, and disposed in panicles at the ends of the branches. It is a distinct addition to the fall bloomers, and so can be used to advantage. It rises from four to six feet. It is counted hardy, though in some latitudes it freezes to the ground, but makes growth enough during the summer following to permit it to be as floriferous as though its stem and branches had withstood the wintry blasts unscathed. In fact, the young foliage is more vigorous and showy than when produced on the last year's growth. *S. sorbifolia* is another form with pinnate leaves, and sessile leaflets lanceolate and doubly as well as sharply serrated. It, too, has large white flowers in terminal panicles, but they appear in July and August, a month earlier than the preceding. It is a native of Siberia.

S. arguta is a more recent introduction from Japan, and is closely related to the *thunbergii* of the same country. It grows about three feet in height, with numerous slender branches, forming a rather open head with small, deep green leaves. The flowers are small, pearly white, and in such abundance as to envelop the whole bush. These break out in very early spring, appearing in April or May,

Spiræa. 133

according to location. The plant is entirely hardy throughout the Northern States, and is known to thrive far south, and westward to the Rocky Mountains. Some authorities pronounce it the best of all the smaller spiræas as a spring bloomer, and it is certainly an elegant shrub,

SPIRÆA ARGUTA.

and as such, when better known, is sure to occupy a prominent place in garden cultivation.

There is a multitude of additional species and varieties, each with its special merits, but so closely resembling one or more of those already described that the differences are of slight significance in practical application. One can scarcely go amiss in making selections, keeping in view what is wanted as to size and season of flowering.

XANTHOCERAS.

Xanthoceras sorbifolia is a small tree but little known among horticulturists and gardeners, though it has been long enough in the country to have gained a much wider distribution had its merits been fully appreciated. It is a native of China, and is the only species of its type, of the order *Sapindaceæ*. The name comes from two words, *xanthos*, signifying yellow, and *keras*, horn, and is applied because of the peculiar horn-like glands or nectaries between the petals. It is said in its native country to form a tree in some instances twelve to fifteen feet high, but in American gardens the few specimens known have reached little more than half those proportions. What they may do in the future in this direction it is not easy to determine. The leaves are alternate, compound, and serrate, resembling those of the mountain ash, while the flowers are white with blood-red streaks at the base, having five petals and eight stamens. They are produced in simple racemes at the end of the branches, the individual flowers being about an inch in diameter. These are not only very attractive, but they are followed by a three-celled fruit said to be "of the size of an apple," which, considering the different sizes to which apples grow, is not very definite.

This shrub, instead of being new, was first pictured and described in the London *Garden* as long ago as 1875, and it has been more or less in cultivation in Europe aud America ever since. It has been grown on the estate of Charles A. Dana at Dorosis, Long Island, for a dozen or more years without especial protection, and though not regarded a strong growing plant, its delicacy of habit is

XANTHOCERAS SORBIFOLIA.

esteemed one of its peculiar charms. Mr. E. S. Carman reports growing the plant from seed on his experimental grounds in New Jersey, and, having seen his original specimen in blossom, says the flower-clusters resemble somewhat those of the horse-chestnut, having "white petals marked with red at the base." As they appear in early spring they are certain to answer a good purpose in the garden, on the lawn, or wherever else planted. It ought to be a valuable acquisition for the more Southern States, and doubtless will be so regarded as soon as better known in that section. It may be depended upon, so far as weather conditions are concerned, in all proper situations south of Washington.

ILEX—Holly.

THE genus Ilex belongs to the order *Ilicineæ*, and includes numerous species and varieties. These are distributed throughout both hemispheres, and are especially abundant in South America and within the tropics. They are also found in Australia, Africa, Asia, and our own country. Of course, many of them are not suited to out-of-door cultivation within the temperate zones, and so need not here be even named or further referred to. The family includes the holly, the *prinos*, and some other sorts popularly known under different names. Nearly all the hardy species are interesting plants, and some of them are counted among the most desirable of ornamental trees.

I. aquifolium is the well-known English holly, a native not only of Britain and other portions of Europe, but also

Ilex—Holly.

of western Asia. It grows in the form of a small tree, ten to twenty feet high, but sometimes reaches much larger proportions, and is famous for its small, round red berries and its glossy, prickly foliage. The leaves are oblong-ovate, deep green, wavy, sharply toothed, and very glossy. The flowers appear in June, and the fruit in late autumn, the berries continuing well into winter. Though perfectly hardy in most parts of England, it is not so in our Northern States. Still, fine specimens are to be found in gardens as far north as the Middle States, and, under favoring conditions, in New England. The species has been prolific in varieties, many of which are more beautiful than the original, and are worthy of general introduction to such portions of America as are fitted to receive and care for them. None of them will withstand our extreme northern winters, though many may be grown under glass to advantage, and will repay such treatment. Nicholson, in his *Dictionary of Gardening*, gives a list of these varieties, some of the best of which are as follows : *I. a. balearica* is held to be one of the best of its class. It is pretty well known in certain sections of the United States, where it thrives, as having ovate and exceedingly shiny black, entire or spiny-toothed foliage. It is supposed to have originated in Minorca, and is often known as the Minorca holly. *I. a. crassifolia* has dull green, very thick leaves with recurved margins and prominent saw-like teeth and purple bark. It is a dwarf and a slow grower. *I. a. doningtonensis* has lanceolate leaves often turned to one side so as to become sickle-shaped. It has few or no spines, and is especially adapted to pyramidal rows. *I. a.*

Ornamental Shrubs.

hastata appears with leaves from three quarters to one half inch long, and about half an inch broad. The spines are very large, consisting usually of one or two pairs on each side of the base, but occasionally more. It is one of the most remarkable forms which the plant puts on. *I. a.*

ILEX AQUIFOLIUM AND OPACA.

myrtifolia has ovate-lanceolate leaves, broad and usually spiny at the margins, but sometimes quite entire; known also as *angustifolia*. *I. a. platyphylla*, as its name implies, has broadly-ovate leaves, having spines sometimes at their edges. It is a hardy evergreen from the Canary Islands, and grows in pyramidal form. *I. a. whittingtonensis* is pronounced an elegant and distinct form with lanceolate leaves two and a half inches long and one and a half wide, slightly recurved, and with many stiff spines.

There are also numerous varieties with gold- and silver-leaved foliage, that are still more showy. Among these is *I. a. argentea marginata* which has broadly-ovate leaves,

Ilex—Holly.

dark green with the disk slightly mottled, and with an irregular, narrow, silvery margin. *I. a. argentea elegantissima* is much like it except that the central part of the leaf is dark green with gray blotches, and has a margin of creamy white. The foliage of *I. a. argentea medio-picta* is dark green at the edges with a large central blotch of creamy white. *I. a. aurea latifolia* is also a broad-leaved variety with well-developed spines, the disk marked with pale green, and having a narrow, irregular, golden edge. *I. a. aurea maculata* has golden-spotted foliage, and a leaf-blade with creamy-yellow centre surrounded with a creamy-white border. It is a very distinct variety and very showy. *I. a. aurea picta latifolia*, popularly known as the "golden milk-maid," has long leaves with spines variable in condition, and disk irregularly marked by a long yellow blotch, with irregular, narrow, glossy margin. This is one of the very best. *I. a. aurea regina*, known as the "golden queen," is a variety in which the leaves are usually much mottled with gray and green, with a broad, well-defined, continuous margin of deep yellow. This is a handsome form and claimed to be the finest of the golden-edged series. *I. a. ferox argentea*, known as the "silver-striped hedgehog" holly, has deep green leaves bristling with stiff spines towards the front and edges. The margins are creamy white. *I. a. handsworthensis* shows much longer leaves than most others, and they are bordered with very strong white spines. The disk is mottled with green, and there is a distinct margin of creamy white. *I. a. wateriana* has leaves oblong-ovate, and disk of dark green, mottled often in sectional streaks with yellowish-green and grayish-green, with broad but irregular marginal bands of deep golden yellow, not continuous, being sometimes wholly golden and at others partly so. This is a dwarf shrub and very beautiful. There are many others almost or quite equally good, but this list will certainly answer all practical purposes.

As already indicated, there are numerous species of the ilex which are natives of Japan, and known to be among the most interesting and beautiful members of the family. The best descriptions of these are furnished by Prof. Sargent, first published in *Garden and Forest*, and later incorporated in his book on the *Flora of Japan*. The descriptions of those which follow are largely compiled from that excellent treatise; and I take this method of according the distinguished author due credit.

I. latifolia is declared by Prof. Sargent to be the most beautiful of all the Japanese hollies, though there are a much greater number indigenous to that country than to any other, for while America has but four species Japan has no less than eleven. *I. latifolia* as known in this country is but a good-sized shrub, while in the land of its nativity it sometimes grows into a tree from fifty to sixty feet high. It may not have been long enough with us to become fully developed, but it is doubted if in this climate it will ever be seen in such proportions. The foliage is especially attractive, the leaves being about six inches long and three or four wide, very thick, dark green, and exceedingly lustrous. The fruit is described as brilliant scarlet and ripening in late autumn or early

Ilex—Holly.

winter. It is produced in axiliary clusters and continues on the branches until the following summer. Prof. Sargent concludes his description by saying: "*Ilex latifolia* is probably the handsomest broad-leaved evergreen that grows in the forests of Japan, not only on account of its brilliant fruit but also on account of the size and character of its foliage. It may be expected to prove hardy in Washington, and will certainly flourish in the southern Atlantic and Gulf States."

I. integra is another beautiful and desirable plant, which has been introduced here where it is sometimes seen as a small tree, but oftener as a mere shrub. It is already recognized as one of the most desirable of the newer sorts now coming into use. It is said to be often planted in the temple gardens of Japan along with *latifolia*, and highly appreciated wherever known. It is not quite as free a grower as the preceding, but is scarcely less desirable. The leaves are obovate, three or four inches long, narrow, with entire edges, and continuing during the winter. The fruit is about half an inch in diameter, abundant, and holding until the next season, and at all times very showy. A variety, known as *I. leuoclada*, is a northern form, and proves to be a dwarf but two or three feet high. This is described as having narrower leaves, and smaller fruit, and will doubtless prove hardier than most of the other Japanese sorts in this country. It is practically unknown, as yet, in American horticulture, but gives promise of becoming a desirable acquisition, especially for northern planting.

I. crenata is better known to us than either *latifolia*

or *integra*, and can be readily procured from many of our best nurseries. It may not have become very widely distributed, but it is a gem worthy of much more consideration than it has apparently yet received. It is a low, much-branched, and somewhat spreading shrub three to four feet high, but in cultivation it not infrequently rises to a height of twenty feet, and so assumes the habit of a tree not unlike the box in general appearance. The leaves are light green, scarcely more than an inch long, ovate with pointed apex and finely toothed. The fruit is black, and produced in great quantities, and contrasts well with the foliage. "This," says our authority, "is the most popular of all the hollies with the Japanese, and a plant cut into fantastic shape is found in nearly every garden. Varieties with variegated leaves are common and much esteemed. *Ilex crenata* and several of its varieties with variegated foliage were introduced into western gardens many years ago, and are occasionally cultivated, although the value of this plant as an under-shrub appears to be hardly known or appreciated outside of Japan. Of the broad-leaved Japan evergreens, I have the most hope of success with *Ilex crenata* in this climate; and if it proves really hardy it will be a most useful addition to our shrubberies." This estimate was made several years ago, and the trials since indicate that it is as great an acquisition as was at that time anticipated.

I. opaca, American holly, is a species, though not so beautiful as the English, that is much to be preferred for planting throughout the North. It grows to about the same dimensions as *crenata*, has oval, flat, deep-green

Ilex—Holly.

leaves, the wavy margins being armed with strong, sharp spines. It, too, is an evergreen, and attractive all the year. The small white flowers appear in loose clusters along the base of the young branches in May and June, though they are never very conspicuous. The fruit is a small, bright red berry continuing on the branches until almost spring. The species is widely distributed along the seacoast from New England southward, but not very plentifully except in a few locations, and then it grows in swamps where it is partially protected from the hot summer's sun and the sharp winds of winter. It is more plentiful southward, extending even to Florida, and again through the barren sections of the lower Mississippi valley. It is accounted difficult of removal for transplanting, and so it is after having acquired age and considerable proportions. But grown in the nursery and frequently transplanted, it may be as safely transferred from one location to another as are most other trees. While it can be made to thrive in almost any good garden soil, it will do better in moist locations and where protected by buildings or trees from piercing winds.

I. verticilata or *prinos* is a native of this country, a deciduous shrub, growing about six feet, whose merits are by no means appreciated. It is sometimes known as the black alder, and in some sections as the winterberry. It has ovate, wedge-shaped, pointed leaves, somewhat serater, and downy on the veins beneath, but its chief excellence consists in the fact that in autumn it is covered with a multitude of crimson-scarlet berries, which hold their places long after the foliage is gone. It is easily one of

Ornamental Shrubs.

the most valuable shrubs that we have for early winter effects, though not especially attractive in summer.

SYMPLOCOS.

THE symplocos group constitutes a large genus of the order *Styracaceæ*, numbering nearly, or quite, one hundred and fifty species. They are mostly natives of warm climates, and in the temperate zones are better adapted to the conservatory or greenhouse than to out-of-door cultivation. None of them is sufficiently hardy to endure the cold of New England or the Northwest, but several evidently have a horticultural mission in our southern latitudes, where they are already more or less planted, and with excellent results. It is now believed that they have also a more extended climatic range northward than has been heretofore supposed, and Mr. Joseph Meehan certifies to the fact that they are growing in the vicinity of Philadelphia, and, under favorable conditions, proving very satisfactory. Specimens are also to be found in southern New England and in New York, which have withstood several winters with but slight extra care or protection.

S. cratæoides, so far as tested, appears to be the most hardy member of the genus, and it is this species that on trial has afforded the satisfactory results referred to. This symplocos is a small tree or shrub five to eight feet tall, and in its general outline somewhat resembles the hawthorn, though differing materially from it in both blossom and fruit. Its leaves are opposite, two and one half inches long, rough and thick. The blossoms are

Symplocos.

small, white, and borne in panicles about five inches in length. They are followed by an abundance of bright ultramarine-blue fruit which ripens in September and constitutes one of the chief attractions. The plant is certain to become a favorite wherever it can be grown, and deservedly so. It is a native of Japan, and is also found growing freely among the Himalayas.

S. tinctoria is a native of the southern United States, where it is popularly called sweet-leaf, because of the fragrance of its foliage; and for the same reason it is sometimes known as horse-sugar. In its favorite haunts it is to be classed as an evergreen, but it may not be found such when carried to the extreme limit of its northern endurance. The flowers are yellow, borne in clusters of from six to fourteen, and exhale an agreeable odor. The leaves are long and narrow, from three to five inches, somewhat coriaceous and sharply toothed. The symplocos grows from three to five feet in height, and is one of our prettiest American shrubs. It can scarcely be planted successfully in the North, but has a field of its own in the lower tiers of States. Treated as is the hydrangea *hortensis* and cared for in winter, it may be grown in the same latitudes.

S. decora is peculiarly adapted to southern cultivation. It comes from China, and is a small camellia-like tree with thick, leathery foliage of considerable beauty. Its flowers are small but abundant, produced in axillary clusters along the young branches. Mr. A. B. Westland says in *Garden and Forest*, in calling attention to this plant: " The petals are white and sometimes tinged with a delicate shade of azure-blue; the cup of the flower is filled with a free cluster of slender stamens each crowned with a pale yellow anther. The size of the flower varies from one half to three quarters of an inch in diameter, and the slightest globular clusters are from three to four inches across. In early spring it bursts into a profusion of delicate blossoms that are gracefully blended with the glossy, green leaves. Its indescribable lightness and grace, combined with its delicious fragrance, make it especially charming."

STYRAX.

THIS genus of the natural order *Styracaceæ*, contains a large number of species widely distributed, a few only of which are sufficiently hardy for out-of-door cultivation, except in tropical or semitropical climates. It is said that the Greek name, by which it is still known, was given to it by Theophrastus, and that in those early days it was even more highly esteemed than now. It was then, and still is, regarded as valuable in medicine, as it produces a balsam, known as storax, highly prized, and yet in use. One of the varieties furnishes what is known as benzoin, but this is not adapted to garden cultivation in this climate. Three species only are natives of North America, and these are found to be somewhat closely related to the halesias, which have been already described. Of these but one is probably worthy of cultivation as ornamental, and even this is surpassed in interest by the introductions from abroad.

S. grandiflora is a small bush growing to a height of from five to seven feet. Its foliage is long and pointed,

STYRAX OBASSIA.

and larger than in most of the other sorts. The flowers are pure white and quite numerous, appearing in early summer, and making it a handsome bush almost certain to attract attention either singly or in the border. It may not be perfectly hardy in the Northern States, but over a large section of the country it can be planted with good results. *S. obassia* is also a shrub of dwarfish habit, with leaves somewhat like those of the catalpa, and racemes of white flowers six inches in length and much resembling those of the well-known mock-orange.

S. japonica, or perhaps more correctly, *S. serrulata*, is held to be superior to the American species above described. Though known here for some years, it has not been grown freely in the nurseries, and so has not been widely distributed. But it is, in fact, a very choice shrub or tree, for it may be grown as either, and, whether one or the other, is symmetrical in shape. If left to take its own course, it usually grows with a single, straight stem, branching low and quite freely. The main branches stand out almost horizontally from the stem, while the branchlets are small, twiggy, and quite numerous. By proper training and cutting out, the tree may be made to take on much the form of a linden or maple, and rise to a height of twenty or thirty feet. But for best floral effects the small twigs should be preserved, as the flowers break out on the whole length of these, and thus almost completely cover the entire framework. They are pure white, about one inch in diameter, and set off by rich yellow stamens. The leaves are small, serrate, sharp-pointed, and rather light green. The plant blooms in early summer, the fruit

Symplocos. 149

following in autumn and hanging in round balls, the seeds in which somewhat resemble kernels of coffee. The plant blossoms and bears fruit when quite young, and is at-

STYRAX JAPONICA.

tractive at all seasons of the year. It has proved hardy both North and South, and thrives in any good garden soil. Mr. Falconer in his notes says that "when in full bloom it is the loveliest plant in our collection, but, alas! it lasts only a few days in flower."

DIERVILLA—Weigela.

ALTHOUGH this group is classified by the botanists as *Diervilla*, the plants are so much better known as weigelas that the continued use of the name under which they were introduced to English horticulturists is still maintained in most of the catalogues, and is perhaps still to be preferred for common use. That name was given by Thunberg in honor of Weigel, a German scientist enjoying at the time considerable distinction as a botanist. But as a French surgeon, Dr. Dierville, had previously reported the discovery of an American member of the family, which, by the by, proves to be the only one indigenous to this country, the genus had been named in honor of this discoverer before the introduction of the Asiatic species to European gardens, and under the well-known law of priority, the name still adheres. The Chinese plant was discovered by Robert Fortune in 1844, and was esteemed by him one of the most beautiful of all the plants which he had been able to gather and send to European gardens from that floriferous country. The first specimen which he saw is described as growing in a Mandarin's garden on the island of Chusan, and characterized as a bush covered with rose-colored flowers, which hung in graceful bunches from the axils of the leaves and the ends of the branches. "Everyone saw and admired the beautiful weigela. I immediately marked it as one of the finest plants in northern China and determined to send plants of it home in every ship until I should hear of its safe arrival. It forms a neat bush, not unlike a syringa in habit, deciduous in winter and flowering in the months

Diervilla—Weigela.

of April and May. One great recommendation to it, is that it is a plant of the easiest cultivation. Cuttings readily strike any time during the winter and spring months, with ordinary attention, and the plant itself grows well in any good soil. It should be grown as it is in China, not tied up in that formal, unnatural way in which we see plants brought to our exhibitions, but a main stem or two chosen for leaders, and then when the plant comes into bloom, the branches are loaded with beautiful flowers which hang down in graceful and natural festoons."

WEIGELA ROSEA.

D. rosea is the plant which thus attracted Mr. Fortune's attention, and is still the best known of the several species. His account, as given above, is sufficiently full and accurate to represent it as it appears in this country, where it has made itself entirely at home. The shrub possesses a tendency to a somewhat straggling growth not altogether objectionable, though it must be cut back severely and at the proper time, if a more regular and compact head is desired. It grows to a height of six to eight feet, with numerous slender stems and branches. The leaves are ovate-lanceolate with finely toothed edges, and are of good color throughout the summer. The flowers put forth in early spring in great profusion, and are deep rose, sometimes freely marked with white. There are several varieties, one, *D. r. nana*, a veritable dwarf with a well-formed, compact head, and a free bloomer. Another, *D. r. nana aurea*, has foliage of a rich golden color, especially in spring during the freshness of the leaves. A third, *D. r. stelzneri*, is distinguished by its multitude of flowers of a reddish-purple tinge. Each of these dwarf forms, of two to three feet, is well suited to crowded situations where there is no room for the larger kinds.

D. candida differs from the preceding chiefly in having creamy-white flowers which do not appear until the first of June. The plant is of a more upright growth and with less disposition to spread by either its roots or branches, and can be used to advantage as affording agreeable contrasts with the darker shades or when planted by itself wherever a shrub with beautiful white flowers at that season is desired. It is greatly to be preferred to the *D. hortensis nivea*, also producing white flowers, though the latter has been much praised.

WEIGELA VARIEGATA.

D. amabilis, lovely weigela, differs from the *rosea* in being of more robust habit and in growing to a larger size. Its blossoms appear later, and at a time when few shrubs in blossom grace the landscape. The foliage is somewhat coarse and the flowers very conspicuous. Many consider it the best of all the family, but it is scarcely entitled to that pre-eminence. Its varieties worthy of mention are the *isolinæ*, having flowers white with yellow at the base ; the *van houttei*, red and white, and the *striata*, red and white in bands. What is known as the variegated weigela, *D. variegata*, is a variety, some say, of the *rosea*, and others of the *amabilis*. It is of smaller size than either, and grows more compactly. The flowers are bright pink and rose, appearing in May or early June. The leaves are beautifully variegated, the margins being creamy-white, and, when well grown, sure to attract attention. As single well-rounded specimens on the lawn, few plants are more attractive than this, and it is equally fitted for massing or ornamental hedging. Though thought to be not quite so hardy as some, it is sufficiently so for all practical purposes, except in the most exposed situations. The white-flowered weigela, *hortensis nivea*, has large, pure-white blossoms which remain long on the branches. The ovate leaves are also larger than most others. It is not entirely hardy in exposed situations. There is a variety whose flowers are deep red when partially expanded, but afterward fade into white.

D. floribunda has blossoms of rich, dark crimson, and somewhat in the form of fuchsias. It comes to us from Japan, and proves perfectly hardy. It blooms abundantly in spring, and, if closely cut back, makes a vigorous growth and puts forth a second harvest of flowers in autumn. Its foliage is dark colored, and contrasts finely with other sorts. It is known also as the *D. multiflora*.

D. arborea is larger than any of the preceding, and easier grown into tree form. Its leaves are large, flowers tube-shaped, much like some of the honeysuckles, mostly pale yellow or rose, and appear after those of most of the other species and varieties have passed by. It is a valuable member of the family, and fills an important place in cultivation. Another good plant is *D. lemoinei*, which produces changeable flowers as to color, pale red turning to deep rose, and again to a rich wine-color. It is esteemed a choice plant, but is not largely cultivated.

Professor Sargent, who during one of his visits to Japan made a study of the wild types, says that in the central and northern sections diervilla, weigela, is a common shrub on the borders of mountain woods and by the banks of mountain streams, and he became of the opinion that what had been referred to by other botanists as several distinct species, are in reality one and the same with variations such as might be expected from differences in soils and exposures. From seeds which he gathered, specimens have been grown in the Arnold Arboretum, and these are known as *Diervilla japonica*. He illustrated them in *Garden and Forest* with the following accompanying description: "It has ovate, acute, or acuminate leaves which are nearly glabrous with the exception of a few hairs on the lower surface of the midribs and veins, or on some individuals these are clothed more or less thickly with soft pubescence. The flowers are borne in few or many-flowered clusters which are long-stalked or nearly sessile, the two forms appearing on the same plant; and they are rose-colored, pale yellow, pale red, or nearly white on the same branch or on different branches of the same plant, and flowers which are pale when they open often become rose-color in fading." When this descrip-

Ribes—Flowering Currant.   155

tion was written, the hardiness of the species for this latitude had not been determined, nor is it yet thought the indications are favorable for its future usefulness.

RIBES—Flowering Currant.

THE currants should not be overlooked among the ornamental shrubs, as some of them prove of special value in garden planting. They belong to the order *Saxifragaceæ*, and are included in the genus *ribes*, numbering between fifty and sixty species. The origin of the name is said to be Arabic, and specimens are found indigenous to Europe, Asia, Africa, and North and South America, growing most freely in mountainous regions, and often at considerable elevations. All are of easy cultivation, and many of them are prized for their fruit as well as their flowers. The foliage of some of the species is liable to mildew, and all are a prey to certain insect pests. These are, however, under modern appliances, so far subject to the control of the gardener as to prove but slight obstacles to success.

R. alpinum is of dwarfish habit, seldom rising above two or three feet, and, as its name indicates, has its home in the mountains. It produces its pale yellow or nearly white blossoms in May. They are in erect racemes, and followed by deep scarlet and very showy fruit. *R. aureum* is the well-known Missouri or Buffalo currant, and is probably more planted than any other. As it was found growing freely on the western prairies and among the foothills of the Rocky Mountains, it took the local names by which it is still popularly known. The bush is larger than the *alpinum*, attaining a height of four to eight feet under favorable conditions, and a breadth of from three to six. Its leaves are three-lobed, toothed, ciliated at the base, and of good color. The blossoms appear in May, and are bright yellow with pink stamens, the petals being considerably shorter than the calyx segments. The fruit ripens late in midsummer, and is also yellow, though occasionally tinted with purple or black. As it is edible, it serves the double purpose of being both useful and ornamental. *R. fragrans* is a variety with larger and more fragrant flowers.

R. floridum is our common black currant, and by some is supposed to be a variety of the *aureum*. It is a native of New England, and grows freely along the Alleghany ranges and throughout the far West. Its foliage is often sprinkled with white resinous dots, and in autumn assumes a tint of bronze which adds to its attractions. The tubular, bell-shaped flowers show themselves in June, and are produced in quite large racemes somewhat downy and of a greenish-white color. The fruit is nearly round, dark-colored, and by many esteemed for culinary purposes.

R. sanguineum grows from three to four feet, and, unlike most of the others, blossoms in midsummer. The flowers are carmine and yellow, and in rich clusters hanging among the leaves and branches, producing a fine effect. There is a double-flowering variety still more attractive, but, unfortunately, neither of these is entirely hardy in the Northern States, and if planted there must be given favored location, or receive especial protection in winter. Farther south it is hardy. *R. gordonianum* is supposed to

Stuartia.

be a hybrid between the *aureum* and the *sanguineum*, and partakes of the good qualities of both. It is the most profuse bloomer in the list, the flowers appearing in hanging racemes of crimson and yellow early in June.

STUARTIA.

THIS is a genus of beautiful shrubs or small trees, containing but a few species, of which two are natives of the United States, and perhaps two or three of Japan and China. It was named in honor of Lord Bute—John Stuart—who gave considerable attention to shrubs and trees, and who is described by a writer in the time of Linnæus as "a most knowing botanist." None of the species are entirely hardy in northern latitudes, but it is proved by more recent trials that they thrive in southern New England, and are as well suited to that fickle climate as are many of our most common shrubs. Of their desirability in garden cultivation there can be no dispute, and Nicholson in his dictionary says they merit a place in every collection of ornamental shrubs. They belong to the order *Ternströmiaceæ*, and produce large camellialike flowers of six sepals and five petals, with a multitude of stamens. All the species should be planted in sheltered positions, as they are unfavorably affected by high winds though seldom suffering from severe cold.

S. pentagyna.—This is a native of the Alleghany Mountain region, extending from northern Virginia southward, and it is also found on the foothills of the Big Smoky Mountains in Tennessee. It is a shrub growing eight to twelve feet, has oval, sharply pointed foliage, and creamy-white flowers in July and August. These are both interesting and beautiful. "The buds are round like those of a peony, swelling to an inch in diameter before the leaves unfold. The petals are one and one-half inches broad in the middle, and two inches long—the flower, borne on a short, strong peduncle, being fully four inches in diameter.

STUARTIA PENTAGYNA.

About the edges the petals are crimped, reminding one of a fluted shell like that of the scallop, and suggesting shell flower as an appropriate familiar name. In the centre of the flower is a cluster of a hundred stamens or more, with prominent, orange-colored anthers. The petals are nominally five, but often we find two or three more, the stamens having changed to petals." When in full blossom the plant is one of surprising beauty, and can never fail to command admiration.

S. virginica is also a native of the State whose name it bears, though it may have grown farther north also, as one of the names by which it was introduced to the public

STUARTIA VIRGINICA.

was *S. marylandica*, indicating Maryland as its habitat. The plant is somewhat smaller than the preceding, and blossoms a month or more earlier. The flowers are white, with purple filaments in striking contrast with their surroundings. There is usually but one style, whereas in the other there are five. The foliage differs in that the leaves are more oblong, serrulate, and downy on the under side. Both have been introduced to European gardens, and received with favor by all who have become acquainted with their characteristics. They do best in a peaty or sandy soil.

S. pseudo-camellia is so named from the fact that its flowers very much resemble those of the camellia. Professor Sargent, in his notes on the forest flora of Japan, says that *Stuartia pseudo-camellia* is common in the Hakone and Nikko Mountains between 2000 and 3000 feet above the sea, where it is a most striking object, from the peculiar appearance of the bark; this is light red, very smooth, and peels off in small flakes like that of the crape myrtle. It becomes there a tree of considerable size; and on the shores of Lake Chuzenji he measured a specimen whose trunk at three feet from the ground girted six feet, and which was upward of fifty feet high. The flowers resemble a single white camellia, are smaller and less beautiful than the flowers of our coast species, *S. virginica*, but are larger than those of *pentagyna*. Specimens were sent to America nearly thirty years ago by Mr. Thomas Hogg, and the tree appears to have flowered in the neighborhood of New York several years before it was known in Europe, where of late it has attracted considerable attention. As known in England and America this plant grows from ten to twelve feet high, with fine foliage, oval, dentate, sharp-pointed at the apex, and narrowed at the base and tinged with red, as are also the sepals of the creamy flowers. When in full flower the entire shrub is often covered with these large, showy blossoms, but it is not certain that even at its best it is superior to the American sorts.

RHODODENDRON.

THE rhododendrons are among the very best of the broad-leaved evergreens, and are everywhere much admired. A large group of these plants when in full bloom is a sight seldom to be forgotten. And in winter they are equally prominent because of their fine foliage. More than almost any other species they are attractive at all seasons of the year, though they cannot always be seen at their best during the colder months because of the partial protection afforded them, and supposed to be necessary, against the extremes of heat and cold.

But, though every one admires the rhododendrons, very few, comparatively, proceed to grow them. The prevailing opinion is that these shrubs are not suited to ordinary cultivation, except within very narrow territorial limits. The failures have been so many and so disastrous that it is thought to be scarcely worth while to continue the experiments, unless it be by experts and under the most favoring conditions. And so, many of the common people pass them by under the impression that though pre-eminently good plants, they are not for them to enjoy in their own gardens and at their own homes.

Ornamental Shrubs.

Still, as a matter of fact, the rhododendrons may be successfully and easily grown over a large section of the country. It is true they may require somewhat especial conditions and intelligent consideration, but even with that they are worth all they cost. To begin with, care must be taken in the selection of species and varieties. This is a clear case where the right plant must be in the right place. The pedigree of each specimen must be thoroughly studied and understood before it can be determined whether it is likely to possess the constitutional qualities required for the purposes to which it is to be put. It may not be necessary for each planter to extend his personal investigations so as to cover all the details in this direction, but the least he can do is to make sure that the nurseryman of whom he makes his purchase is sufficiently intelligent and sufficiently honest to be able to present a clean bill of health at every point and extending to every particular.

Very few of the rhododendrons in ordinary cultivation in this country are to be found growing wild in any part of the world. They are crosses, hybrids, or varieties largely fashioned by the hand of man, and the experts who have toiled long and with intelligent purpose have brought out a race of these beautiful plants such as the world knew little or nothing of before. A brief account of how this has been done may not only be of interest to the general reader, but prove also of practical value to such as wish to grow them on their lawns or in their gardens. Let us, then, first look to the sources from which our common sorts have mostly sprung, and take a glimpse of the proc-

Rhododendron. 163

esses of development from the native forms to those which we now so highly prize. Among the original species which have been brought into reputation, but three of the most prominent need be named.

R. arboreum is the largest known species, and is sometimes called the tree rhododendron in allusion to its size and form. It is a native of the Himalayan Mountains, where it reaches the height of from twenty-five to thirty-five feet, and sometimes, according to Nicholson, acquires a circumference of 150 feet. The flowers are described as white, rose, and blood color, disposed in dense heads and very beautiful. The foliage is equally bold and showy, "the leaves being large, coriaceous, lanceolate, acute, cordate at base, or attenuated into a thick petiole, of a beautiful green above, below impressed with netted veins, glabrous, silvery, or ferruginous-pubescent." Attention is called to this species though it is known to be too tender for out-of-door planting in this climate, except perhaps, in our Southern States, and even then it is not recommended for ordinary cultivation.

Another of the foreign species is *R. ponticum*, which also proves one of the most desirable members of the family, and which, though hardier than the *arboreum*, is yet too tender for our use. It is a smaller plant, growing six to ten feet, with good foliage and exquisitely beautiful flowers. This is a native of Asia Minor, and while hardy throughout most of continental Europe, it proves a practical failure in American cultivation, at least as far north as New York. Though doing fairly well in England, the extremes of heat and cold, and more especially of drought and moisture, forbid its introduction, together with most of its varieties, for our ordinary garden use. Just here comes in one of the secrets of the failure of rhododendron planting in America, so far as it is a failure among the common people. To make sure of the splendid colors of the two species, nearly all the experiments in hybridizing and crossing have been with these comparatively tender plants for the foundation. As the *ponticum* proves hardy on the continent, it was very natural that the nurserymen of France, Belgium, and Holland, the great feeders of the civilized world in the line of nursery stock, should cling to the notion that, being found hardy at home, it must be strong and vigorous enough for other countries lying within the same range of latitude and supposedly subject to similar climatic conditions. And so they have continued sending us their wares, and too many of our nurserymen have continued the purchase and distribution of the French and Dutch plants, budded on *ponticum* roots, by the hundred thousand a year. It is not too much to say that no rhododendrons on such roots should be accepted as thoroughly reliable in our northern American climate. They may appear to be all right for a few years, starting out well, but with rare exceptions they are certain to fail of the highest perfection unless especially protected and pampered. The rhododendron is naturally a long-lived plant, specimens having been found in its native habitat more than a hundred and fifty years old, and still vigorous.

R. catawbiense is an American species, and grows freely on many of our hillsides and mountains without the slightest protection or care from the hand of man. It is far

Rhododendron.

from being one of the best and most showy species, but is known to be entirely hardy and especially adapted to our soil and climate. It has, therefore, an important mission to perform in becoming the foundation of the many new creations which to us constitute the glory of the rhododendron family, and as such should carry the numerous varieties and hybrids, which, enriched by the blood of the *arboreum* and *ponticum*, constitute the rich fields from which our selections are to be made. In England the best growers use only the *catawbiense* for budding or grafting stocks for the American market, and their plants on *catawbiense*

HYBRID RHODODENDRON.

roots are confessedly the best of all the importations that come to us. There is no good reason why we should not grow for ourselves all the rhododendrons we need, but thus far the home production is exceedingly limited. *R. maximum* is another hardy sort, thriving as far north as New England, which can also be used as stocks with equal safety, and there are perhaps still others.

The rhododendron should not only be well fortified in its essential qualities, so far as the constitution of the plant itself is concerned, but its demands must be recognized for peculiar situations and to some extent especial care.

There are some soils in which it positively refuses to grow. It has no affinity for chalk and lime, at least none sufficient to reconcile it to friendly relations. Neither is clay found to be favorite feeding ground, though, unlike the chalk and lime soils, the difficulty may be overcome, by digging wide and deep, and thus creating practically new conditions by filling in with suitable plant food. A rhododendron bed may be thus successfully maintained in the clay soils, while it cannot be done in situations saturated with lime water, as in the course of time the offensive matter will percolate the new soil as thoroughly as it does the old; and then, however hopefully begun, all prosperity is at an end. It may be put down as an established fact that there are some localities where this charming plant cannot be successfully employed in garden or landscape work, though the nurserymen often say there is not. But in most fairly good soils it is as easy to make the rhododendron grow as it is other choice shrubs which are planted in profusion and almost everywhere found to thrive.

Starting out with vigorous stock, the rhododendrons, to prosper, must be well planted, especially if the best results are desired. They are not so particular, however, as many suppose, for they will sometimes become vigorous and showy shrubs and even small trees, though treated with entire neglect and left to themselves among the grasses and weeds. But to make sure of success it is well to give the best care and make the best provision for them that is possible, as in the case of all other choice species. Any good garden soil, with the exception already indicated,

Rhododendron.

will answer the purpose. The rhododendron makes a multitude of fine, almost hairy, roots, such as do not rapidly penetrate far into the ground, and so the soil should be lighta nd somewhat porous, and in preparing for planting there should be an excavation two and a half or three feet deep, at least, and as many broad, for each plant. This should be filled with chopped turf or rich compost, and, if the soil is especially hard or heavy, a mixture of peat and sand. The best plants I ever saw were put into the ground in this way, with nothing else than common garden and turf substances. Something depends on the situation chosen as well as on the method of the planting. The rhododendron is, in a measure, a shade-loving plant, and this fact should be recognized when fixing upon the location, whether for a group or a single specimen. Sunshine in the morning and at evening, and perhaps with rays darting through the tree-tops, at midday, constitutes an ideal situation. So far as practicable, protection from the prevailing winds, either by buildings or neighboring trees, should be afforded, and with these simple counsels any one can plant and easily maintain a bed of rhododendrons. Once established, the really hardy varieties need little or no protection from cold, even where the thermometer occasionally registers several degrees below zero. They may be partially covered with branches of evergreens in winter, but the advantage of this is in prevention of sun-scald more than for protection from frost or snow. Far more rhododendrons are killed by too much sunshine than too much cold, though by the ordinary grower this fact is scarcely ever appre-

168   Ornamental Shrubs.

ciated. They are all moisture-loving plants and often suffer in seasons of drought, and, though not always showing it at the time, become so weakened that they cannot endure their winter hardships. In such cases the real difficulty is not even suspected by most growers. In dry seasons, or in long absence of rain, water should be applied freely and copiously.

R. maximum, known also as the great laurel, is a valuable plant for ordinary cultivation. It is indigenous as far north as Connecticut and Rhode Island, and often

RHODODENDRON MAXIMUM.

grows to a height of twenty or more feet. It is a profuse bloomer, and has the advantage of putting forth its flowers in July after the blooming season of this class of plants is

Rhododendron.

otherwise over. The flowers are pale rose or pink varying to white with many inter-mixtures of shades, but always attractive. As a single specimen and in tree form, few shrubs surpass it in interest, and it is worthy of much more general cultivation. *R. punctatum* rises but five or six feet, has its blossoms in dense corymbs, rose-colored, somewhat funnel-shaped, and spotted within. These appear in May or June. Like the leaves of the plant itself, they are borne on pedicils covered with viscid globules which are unpleasant to the touch. The shrub affects lofty altitudes, and it is seldom found except among and high up on the mountains.

As there are more than a hundred species and innumerable hybrids and varieties of rhododendrons known in horticulture, it is impracticable to attempt an enumeration of their names, much less of their varying qualities. They can only be discussed on general principles and in groups and classes, leaving details to the judgment and taste of those especially interested. There are but three American species which may be considered of value in this connection,—the *catawbiense*, the *maximum*, and the *punctatum*, to each of which reference has been made. The first-named is found growing freely on the Carolina mountains and contiguous localities. It is usually from three to six feet high with an irregular, spreading head, and oval or oblong leaves rounded at both ends. The flowers are lilac-purple, appearing in July, and not especially beautiful. The hardiness and adaptability to adverse situations is seen in the fact that the plant is sometimes found in moist woods and leaf-mould, and sometimes springing from the seams of rocky ledges fully exposed to the scorching rays of the southern sun, and in neither case suffering harm. As already indicated it is this quality of endurance that gives the plant its chief value as affording a substantial basis for the more tender and more showy sorts.

The following named sorts are reported reliably hardy in New England when grown on *catawbiense* roots, and may be planted with full assurance of success: *Album grandiflorum*, which has been longer known than almost any other sort as competent to withstand our severest winters without protection, is still one of the very best, producing large trusses of pink flowers, later on changing to white; Charles Bagley, flowers, cherry-red; Lady Grey Egerton, silvery-blush with grayish-brown spots; Lady Crosley, pink-salmon; Charles Dickens, red, and one of the best of that color; Lady Armstrong, noted for its foliage as well as blossoms; Kettledrum, rose-colored; Sefton, deep maroon; Alexander Dancer, beautiful red; Old Port, plum-color; John Waterer, dark crimson; *delicatessimum*, blush changing to white, one of the very best, being a late bloomer; Mrs. Miller, rich crimson; *everestianum*, rosy-lilac, and reliable everywhere; Abraham Lincoln, fine rosy-crimson; General Grant, rosy-scarlet; *roseum elegans*, fine rose; *giganteum*, crimson-rose, very large; Minnie, white with saffron or yellow centre; *purpureum*, in several varieties, all purple.

To the above, Mr. H. H. Hunnewell, whose gardens at Wellesley, Mass., are famous for their rhododendrons, and who has experimented on a large scale, writes to me that he has found the additional newer varieties entirely hardy in that locality: C. S. Sargent, F. L. Olmstead, F. L. Ames, Mrs. C. S. Sargent, Mrs. R. G. Shaw, Mrs. N. S. Hunnewell, Mrs. Arthur Hunnewell, Mrs. Charles Thorald, Mrs. J. P. Lade, Mrs. Simpson, Lady Grey Egerton, Countess Normantown, Princess Mary of Cambridge, *maximum wellesianum ;* and that several others give promise of becoming valuable acquisitions, but need further testing. For the practical grower these are not named as superior to many other sorts, but as merely indicating the wide range from which selections may be made.

CORYLUS—Hazelnut Tree.

IN the public mind both the corylus and the hamamelis, in their several species and varieties, are classed as witch-hazels, and spoken of accordingly. But, though having some things in common, botanically considered they are wide apart, and should not bear the same name. Like *Hamamelis virginica*, the American hazelnut, *Corylus americana*, is a native of the western continent, and a shrub of value for planting in masses for screens along water-courses or in other moist localities. The common species constitutes a familiar object throughout a large portion of the country, especially in the Northern States. It is sometimes planted for its fruit, but not often, though it might possibly serve a good purpose in that line. The European hazelnut is scarcely an improvement on the native plant, except that it it has given us two varieties, each of which has a distinct value. One of these has cut-leaved foliage of a somewhat peculiar character, and can be used to advantage in the shrubbery or border. The other, *C. purpurea*, or purple-leaved hazel, is very ornamental, and worthy of a place in almost any collection. Early in the spring the opening buds expand into large, finely formed leaves that are almost black—nearly if not quite the darkest shade known in vegetation. The effect is then exceedingly striking, as the contrast with surrounding objects is very marked. A little later on the color changes to a lighter hue, becoming a dull purple, and thus continuing most of the season. The coloring is much the same as that of the darkest purple beech, which entitles it to a position among shrubs such as that tree has secured in the great family of larger growths. It holds its color much longer than does the beech. The shrub grows from four to six feet, and thrives in dry as well as in moist soils.

SASSAFRAS.

THE sassafras is a native of the eastern portions of the United States, ranging from Canada to Florida, and adapting itself to these extremes of heat and cold. It constitutes a genus of the order *Lauraceæ*, with but a single species and few or no marked varieties. The name, as given by Linnæus, and adopted by about every authority since his time, is *Laurus sassafras*, but modern botanists now propose to call it *Sassafras sassafras*, which leads Dr. E. S. Bartin, in the *American Journal of Pharmacy*, to say that this is "doubtless applied in strict accordance with the new rules for botanical nomenclature ; but whose

Sassafras.

unpleasant effect upon the ear could not be well endured, except in the hope that sometime between now and the millennium our botanical nomenclature will acquire something like a stable equilibrium." The tree never grows to a large size, though sometimes rising to a height of forty to fifty feet, and that, too, when the trunk is scarcely more than one or two feet in diameter. In the extreme North it is little more than a tall slender shrub. The bark of the trunk is somewhat gray in color, and deeply furrowed, but on the young branches the covering is usually green tinged with red. It is impossible to give a description of its leaf that will fit all cases or even apply wholly to a certain tree, for the sassafras has the peculiarity of bearing leaves that do not resemble each other. They are usually about four inches long, petioled and alternate. Some of them on the same stem are oval and entire; some have a rather small lobe on each side; others are lobed on one side and not on the opposite, and still others appear with three lobes. The

SASSAFRAS OFFICINALE.

flower is yellow, not very conspicuous, and appears in advance of the leaves.

The sassafras was one of the first of American trees to attract the attention of Europeans. It was carried to the Old World as early as 1540, and in 1549 a treatise was published by Gerard, who called it the ague tree, and pronounced a decoction of its bark a cure for many diseases. For a long time its real or supposed medicinal virtues gave it a high place among the physicians, and its merits are still recognized. Its virtues in this direction are said to come largely from the inner bark, both of the trunk and roots, which is of a dark reddish color not altogether unlike the celebrated Peruvian bark. The flowers and twigs are also in use, and the wood is sold in chips for medicinal purposes. Few trees or plants have held their reputation so long and through so many vicissitudes as this. A large number of supposed species or varieties are found in a fossil condition,—that is to say, the leaves of such trees are found, and it has been supposed they belonged to different species, from the fact of their varying in character,—but as our present species of sassafras has numerous forms, even on a small tree, it does not follow that these ancient geological specimens are of extinct sorts.

LAGERSTRŒMIA—Crape Myrtle.

THIS constitutes a small genus of the order *Lythraceæ*, consisting of about a dozen species, all natives of Asia, and but little known in cultivation. They are mostly greenhouse plants, two or three only being suf-

Lagerstrœmia—Crape Myrtle. 175

ficiently hardy to endure the climate of any portion of the United States. Of these *L. indica* is best known and most highly appreciated. It grows from six to ten feet, and has somewhat oval leaves, acute and glabrous. The blossoms are large and very beautiful, produced in panicles, are bright pink with the petals curled, and on long

LAGERSTRŒMIA.

claws. They appear in midsummer, and the effect is very striking. The shrub is found to be sufficiently hardy to thrive in the Middle States in protected situations, and is quite at home farther south. There it proves a most desirable acquisition, and is fast coming to be appreciated. It is popularly known as the crape myrtle, a name suggested by a peculiarity of the flowers. There is a variety with white blossoms, making an agreeable contrast when the two are planted in conjunction. Another species, *L. flos-reginæ*, commonly called queen's flower, is equally suited to garden cultivation, and is perhaps even more interesting. The flowers are from two to three inches in diameter, of a beautiful rose-color in the morning, changing gradually to purple at night. The shrub is a native of China.

AMELANCHIER.

THE amelanchiers are well known shrubs or small trees to which more attention should be given in ornamental planting than they have yet received. They belong to the order *Rosaceæ*, and are widely distributed, though until quite recently supposed to be almost exclusively natives of North America. With the opening up of Japan several new species were discovered, some of which prove to be of value, being largely planted in Europe, and to some extent in the United States in competition with those which are natives of the soil. There is some confusion among the botanists as to classification and names, but the questions raised are comparatively of small importance to practical horticulturists, however entertaining and interesting to the experts. The nomenclature in use by Nicholson, Torrey, and Gray will be followed.

A. canadensis is more common both in our forests and gardens than any other species. Loudon found it in abundance in the American woods during his early botanical tours, and first brought it to the attention of the general public, describing it as follows: "A very orna-

Amelanchier.

mental tree, from its profusion of blossoms early in April, and from its rich autumnal foliage ; and even the fruit is not altogether to be despised, either eaten by itself or in tarts, pies, and puddings. The wood is white, and it exhibits no difference between the heart and sap. It is longitudinally traversed by small, bright red vessels, which intersect each other and run together—a physiological peculiarity which, Micheaux observes, occurs also in the red birch." As might be expected of a shrub or tree whose habitat extends from Hudson's Bay to the Gulf States, and from the Atlantic to the base of the Rocky Mountains, and possibly even beyond, it is perfectly hardy and full of vigor. It is sometimes of bushy form, but more frequently appears as a small tree rising from fifteen to thirty feet. Its pure-white flowers are disposed in short racemes, and so numerous are they that the foliage and branches are almost hidden from sight. In autumn it is almost equally beautiful, its foliage taking on bright golden-yellow tints, thus closing as well as opening the season with a show of beauty. The fruit, popularly known as the June berry, service berry, and shad berry, is about the size of a currant, of purple color, and agreeable to the taste, so that where the trees are abundant it is often gathered as an article of food. It matures in June or early July. The tree blossoms and produces fruit while quite young, and it is no uncommon thing for a specimen of but three or four feet to be heavily laden, thus affording quick returns to the planter. *A. botryapium* is given by Gray as a synonym, and by others as the name of a variety.

A. alnifolia is a Western species or variety of much smaller proportions, being a veritable dwarf though possessing all the good qualities imputed to the larger plant. It is of especial value where a smaller form is desired through limitations of space or adaptation to surrounding objects. In some locations it is grown especially for its fruit, which is esteemed above that of the currant, while the bush is much more ornamental, and occupies no more space. It has sometimes been put on the market as a blueberry, but is of quite another family. It has the advantage of producing both blossoms and fruit when not more than two or three feet high. The flowers appear later than those of the larger sort, and so help to lengthen the season. *A. rotundifolia* is much the same except that the bush grows some two feet taller, and there are several others offered in the market with differences so slight as to call for no further description.

A. japonica is a recent introduction to American gardens, and appears to be worthy of notice. It, too, has early blossoms in great abundance, not differing materially from those already described. Its fruit is bright scarlet, and proves very showy, and hangs on a long time, often after the leaves have fallen. It is a strong grower and highly ornamental in its foliage as well as in its fruit and blossom.

DAPHNE.

THE daphnes constitute a very interesting genus of the order *Thymelæaceæ*, the several species of which are widely distributed throughout the south temperate zone, and in smaller numbers in the southern hemisphere. It is said that there are some forty species and varieties of which note has been taken by botanists and horticulturists. They are all small shrubs, though in some cases in tree form, and present numerous attractions. Some are evergreen and others are deciduous. In cultivation a few are known only as greenhouse plants, but the great majority are counted hardy and suitable for garden and park planting.

D. mezereum is one of the very best of early spring bloomers, and merits much more attention than it has yet received from horticulturists and planters in this country. The plant is perfectly hardy, and it is claimed will grow up to the very borders of the Arctic regions. Whether it will endure such an extreme test or not, it is certain that it can be used freely in all portions of the United States short of Alaska, and is also available in the South to the borders of the Gulf. It has been in use in English gardens for more than three hundred years, and has held its own against all newcomers for that long period. It is a low-growing plant, seldom rising more than two feet, but is of a somewhat spreading habit and so covering considerable space. The flowers are pink, very abundant and very fragrant, appearing at the first breath of spring and often before the snow has fully gone. When these are over, the little bush is almost equally attractive from the presence and character of its fruit. In June the branches are crowded with large, bright red berries, affording a marked contrast to the dark green leaves. Coming thus when so few plants are in blossom, and continuing so long in fruit, the wonder is that it is not better known and more largely planted in ordinary gardens. It is not easily propagated in the nursery, and so costs slightly more than some larger sorts, and this may account for a portion of the comparative neglect. But, as already stated, when established in the border or planted in groups it is equal to the best, and will withstand almost any exposure. And it is not a small thing in its favor that it blossoms every year, and is not very particular as to soil or surroundings.

D. cneorum.—In this we have another small shrub of which it is difficult to say too much. It is popularly known as the garland flower, has evergreen foliage, and blooms during May in New England and the North, and much earlier in warmer climes. The flowers are of a rich lilac shade, and profuse in their abundance. During the summer they appear from time to time, and in autumn often break forth anew and cover for the second time the whole bush. A writer in the London *Garden* speaks of a full crop of flowers as late as December, and *Garden and Forest* says that plants in the neighborhood of New York were blooming on the twelfth of the same month, opening their flowers perfectly and giving forth a fragrance which seemed even richer than that of the spring bloom. This plant is especially recommended for rockwork, for borders of shrubberies, and for planting in groups where masses of color are desired.

D. genkwa is a deciduous shrub growing two to three feet in height, with numerous twiggy branches clothed with a soft down. Like all the daphnes, it blossoms in early spring, producing violet-colored flowers the whole length of the otherwise naked branches. They continue two or three weeks, and are quite fragrant though not especially beautiful. Some of the varieties are superior to the type, having larger flowers and of a richer shade of dark purple, while one, not yet known to cultivation, has been discovered with white flowers. Though not the best of the family, the species is worthy of cultivation, and in some situations proves of special value.

D. laureola.—This takes its popular name of spurge-laurel from its foliage; the leaves are green, oblong, and remain through the entire winter. It is a low plant—even smaller than *cneorum*, and less beautiful in flower, as the blossoms are of a greenish cast and without fragrance. But it is a good foliage plant, and has the quality of growing in the shade, and especially under trees, better than most other sorts. For such situations it serves a good purpose, and may be used to advantage as one of the best under-shrubs.

VIBURNUM.

THE viburnums constitute a genus of small trees or shrubs of the natural order *Caprifoliaceæ*, representing about eighty species, mostly distributed throughout the north temperate zone, though a few specimens are also found in the West Indies and Madagascar. Much the greater number are indigenous to America. They usually have opposite branches, with undivided, lobed leaves, white, terminal, tubular flowers, and seed-like berries. They prefer moist places, but grow well in partial shade in any good soil, though preferring a peaty substance to sand or gravel. The more valuable are the following:

The English wayfaring tree, or hobble-bush, *V. lantana*, known also as the rowan tree, is a shrub ten to twenty feet high, with small white flowers, in large, flat cymes, appearing in May or June. The fruit, which remains long on the stem, is a bluish-black berry, somewhat flattened and quite sweet to the taste. The leaves are two to four inches in length, rough or crinkled in appearance, downy, and often with hairs on the under side. The inner bark is acrid, and both it and the fruit have been much used in medicine. It is indigenous to countries as far north as Scotland and Siberia, and widely distributed. The rowan tree was long supposed to be a protection against witches, and in times when men put faith in goblins and spirits intent on mischief its efficacy was seldom doubted; for that reason it was planted in close proximity to dwellings and stables.

The American species, *V. lantanoides*, differs somewhat from the above, as it appears less in tree form and more as a low, bushy shrub. It grows wild in dark, moist, rocky woods as far north as New Brunswick, and thence south through portions of New England to the mountains of North Carolina; and often makes a beautiful show of flowers in unfrequented and desolate places. The leaves are from four to six inches across, shaped much like the English variety, smooth above but downy along the veins. The flowers are in broad heads of white, and are followed by crimson fruit, afterward turning black. Breck says: "The first time we beheld the crooked, straggling shrub in flower in its native haunts, a dark swamp, we thought it one of the most ornamental shrubs in the

Viburnum.   183

country. It is certainly worthy a place in every collection." There is a variety of considerable value, with the same general features, but having variegated leaves white and yellow.

The sweet viburnum, or sheep berry, *V. lentago*, grows from fifteen to thirty feet high, and is valuable for ornamental purposes. The flowers are profuse and showy, appearing in June. They are composed of a large number of terminal cymes, making a broad, white head with a slight tinge of yellow. The fruit consists of rich, dark blue berries, and, as they are in marked contrast to the autumn foliage hues, the appearance is very striking. The leaves are ovate, pointed, and on long, marginal petioles.

VIBURNUM OPULUS.

The so-called high bush cranberry, *V. opulus*, grows from five to ten feet, and is a vigorous shrub showy in both its flowers and its fruit. Emerson says: "In May or early in June it spreads open at the end of every branch a broad cyme of soft, delicate flowers, surrounded by an irregular circle of snow-white stars, scattered, apparently, for show. The fruit, which is red when ripe, is of a pleasant, acid taste, resembling cranberries, for which it is sometimes substituted." This plant is beautiful in flower, in leaf, and in fruit; and as the fruit remains well into the winter, ever deepening in color, the beauty of the bush lasts all the year. It is the parent of the well-known snowball tree, which under cultivation produces only sterile flowers. In this form it has long been well known. There is also a variety with variegated foliage, the leaves marked with yellow and white. The *V. o. nana* is the dwarf of the family, for, though perfect in all its forms, it seldom lifts its head more than a foot from the ground.

The naked viburnum, or wythe rod, *V. nudum*, is also a native of America, and indigenous as far north as New England and southward to Florida, proving entirely hardy in both sections. The flowers are yellowish-white, appearing in May or June. They are composed of small florets in large, crowded heads. The fruit is nearly round, quite large, of a deep blue color, ripening in September or October. The leaves are oblong-oval, with a rather rough surface. It is an interesting shrub, but less valuable than several of the other species.

The Japan snowball, *V. plicatum*, is one of our later acquisitions, and is pronounced by good judges the most beautiful and most desirable of all the members of the family. It is of moderate growth and compact habit. The leaves are crinkled or plicated, of a rich green color, borne on brown shoots, while the flowers are larger and more solid than those of the common snowball. They hang long on the bush, and are very showy. The great-clus-

Viburnum.

tered snowball, *V. macrocephalum*, was first found in the gardens about Chusan, in China. It proves to be the equal of the old American snowball, or guelder rose, in purity of color, and far eclipses it in size and beauty. Each blossom is more than an inch across, and the clusters made up of these measure eight or ten inches

VIBURNUM MACROCEPHALUM.

in diameter. It is sometimes advertised as a new species from Japan, but, though rare, has been planted here for many years. The tree grows to the height of about twenty feet. It is sufficiently hardy to withstand the winters of New England.

Arrow-wood, *V. dentatum*, was popularly so named because the Indians used it in the manufacture of their arrows for use in hunting and in war. The wood is heavy, exceedingly hard and tough, and capable of high polish. The shrub is small, seldom growing more than five or ten feet, with light-colored bark, and pale green leaves sharply dentate. The dark blue flowers are large and showy, folowed by purple fruit that hangs long on the branches. It is an interesting plant for the garden or border, and, though found chiefly in swamps, will grow on ordinary rich land to perfection. Downy arrow-wood, *V. pubescens,* is not as interesting as are most of the other species, and is scarcely found in our northern latitudes, while appearing in abundance in the swamps of Kentucky and farther south. Gray describes it as a low, straggling shrub, having ovate and taper-pointed leaves, with a few coarse teeth and a downy surface. The fruit is dark purple, much like that of *dentatum.* Maple-leaved arrow-wood, *V. accrifolium,* is a low, slender shrub, three to six feet, the leaves of which have a close resemblance to those of the maple, being three-ribbed and three-lobed. It is entirely hardy, and often found growing in the forests of New England. The flowers are borne on terminal cymes, with slender stamens, and are of a pale purple color when first opening, the corolla afterward becoming pure white.

The dilated viburnum, *V. dilatatum,* is a native of Japan, and grows to a height of eight or ten feet. The small flowers are in cymes from two to five inches across, and very pretty, appearing in June. The best authorities pronounce it a valuable shrub, though it does not appear to have been largely planted as compared with *plicatum,* which everywhere now takes the lead. The fragrant viburnum, *V. odoratissimum,* is a shrub from the Chinese

Viburnum.

mountains, and rises to a height of ten feet. The flowers are very sweet-scented. They are in corymbs, white and quite showy in May, being among the first blossoms that put out an appearance. It has the reputation of not being

VIBURNUM TOMENTOSUM.

entirely hardy except in situations protected from high winds and extreme cold. *V. tomentosum* differs from the *plicatum* in having more hairy foliage and less double flowers. It is hardy and desirable.

The evergreen viburnum, *V. tinus*, is a native of

188   Ornamental Shrubs.

southern Europe, and retains its foliage through the winter. The leaves are oblong, entire, and sometimes

VIBURNUM PRUNIFOLIUM.

hairy. The flowers are at first rose-tinted, but soon become pure white. They come forth in late autumn, and often at warm periods in winter from December to March, when they never fail to attract attention. The fruit is dark blue, resembling that of several of the other species. The shrub grows from six to nine feet under favorable conditions. It is a curious plant, and well worthy of attention. The variety known as the *fræbelii* has lighter-colored flowers. Another variety, the *lucidum*, brought from Mount Atlas, has large glossy leaves, and also larger flowers than the original. They appear in early spring.

The wrinkled-leaved viburnum, *V. rugosum*, is an evergreen growing from four to six feet. The flowers are not especially attractive, but the peculiarity of appearing in winter creates an interest in the plant. Its habit is much the same as that of the *tinus*, but it is not so hardy and does not hold its foliage so persistently.

What is known to some as the American black haw is the *V. prunifolium*, named from its prune-shaped leaves. It becomes a large shrub or small tree, but is less valuable than many of the other sorts named; still it is not without interest.

BUXUS—Box.

THE box is a genus of hardy evergreen shrubs or small trees, of the order *Euphorbiaceæ*. These plants were largely in use among the Greeks and Romans, and were highly spoken of by some of the most noted writers of antiquity. The Greek word indicating the character of the shrub signifies strength, or sometimes a cup, and is supposed to have been applied from the use of the wood, which is especially fine-grained and hard, and was largely used in the manufacture of goblets and ornaments. All the members of the family are broad-leaved evergreens, but not all are sufficiently hardy to withstand our extreme northern winters. They are furnished with opposite leaves, entire at the margins, and possessing a peculiarity by which the plates can be easily split. The blossoms are numerous but inconspicuous, growing in axillary clusters, the male and female specimens, though distinct, being on the same plant. There are but two species in ordinary garden cultivation.

B. sempervirens is the most widely distributed, and is commonly known as the tree box, specimens of which may be seen in many of our old gardens. It is nowhere indigenous to America though found in similar latitudes throughout England and southern Europe, and even as far east as Persia and perhaps China. At its best it is said to rise in tree form from twelve to fifteen feet, but, as it advances toward the colder regions, it gradually diminishes to three or four feet. The leaves are small, oblong-oval, bright green, and somewhat coriaceous. The wood is hard, close-grained, and capable of a very high polish, the specific gravity being such that it will not float in water. Though small, it is a long-lived tree, many specimens found in the Eastern States being considerably more than a hundred years old and still in good condition. The species has given off numerous varieties, some of which as garden plants are to be preferred to the type. *B. suffruticosa*, or dwarf box, has been used for edgings to paths and borders more largely than, perhaps, any other plant. Nearly all the old-fashioned gardens in this country, following the custom of Europe, had their box borders, so much so that they became somewhat monotonous, and a revulsion came, resulting to a considerable degree in their discontinuance. But of late the tide has turned, and the tendency now is in favor of their restoration. Another departure is the *B. s. argentea*, or silver-variegated, in which the foliage is marked with white. This, though a larger and more rapidly-growing plant, can be employed to advantage for hedges, or used as single specimens for grouping, and is also capable of good service as an undershrub in parks and other large grounds where partially shade-loving growths are desired. The *B. s. aurea* and the *marginata* have their leaves marked with yellow, and may be used in the same way. Another variation, known in the catalogues as *elegantissima variegata*, is said to be very fine. The *macrophylla* has larger leaves, more oval in form than the type, while in the *microphylla* they are much smaller. *Handsworthii* is distinguished as an upright form with still more oval leaves and as being a vigorous grower. It has deep green foliage, and is credited with being especially hardy, and thriving where most of the others are liable to fail.

B. balearica, known also as the Minorca box, has larger foliage than *sempervirens*, and is also a native of western Europe and Asia. It latterly has come to be called by some the Japanese box, though for no good reason, as Japan is only one of the many countries in which it is indigenous. Its yellowish-green leaves are about two inches long and a little more than half as broad. It sometimes attains a height of fifteen to twenty feet with a large

Ornamental Shrubs.

though compact head. As a single specimen it is probably the best of the family, but, unfortunately, it is not quite so hardy as *sempervirens*, and in northern localities must have winter protection, especially when young. In the Middle and Southern States it may not need extra care. It, too, has gold- and silver-leaved varieties, and some of the handsomest plants sent out from the nurseries are of this class.

The box is capable of being grown in any form desired, and is entirely submissive to the knife or pruning shears. In the days when more formal gardening was in vogue, and plants were cut and grown in fantastic shapes, the box was one of the favorite plants for use in this direction. At the present time that absurd system is almost unknown, though occasionally an attempt is made to copy from the old models as a matter of curiosity rather than the desire of restoring the stiffness of the straight lines and geometric and sculptural figures of what is popularly called Italian gardening.

KOELREUTERIA.

THIS native of China is named for the German professor, Koelreuter. It is of the order *Sapindaceæ*, and is a small tree of particularly picturesque habit of growth, which makes it highly desirable for lawn or garden. On the European continent, where the planting of these dwarf trees is understood to perfection, they are used to an extent unthought of in America, and are placed where effect and, in some cases, shade are desired without interference with view or a free circulation of air. A better understanding here of the subject would prevent the

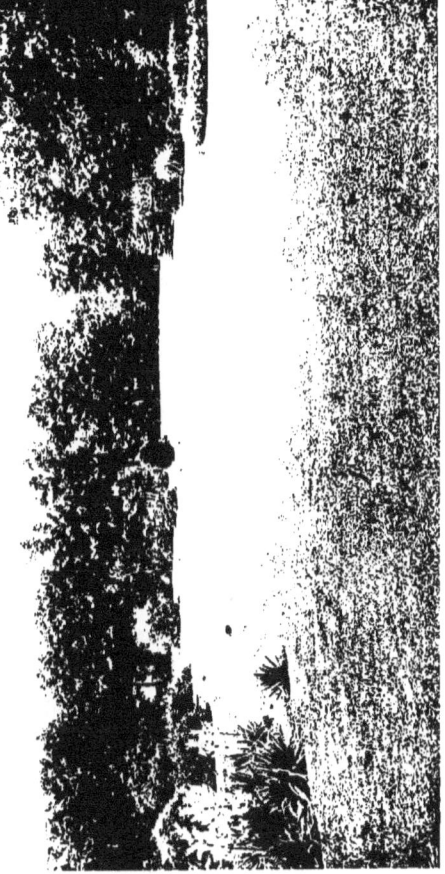

VIEW FROM VILLA OF W. C. EGAN, AT EGANDALE, CHICAGO.

disappointment often experienced by the owner of a small place, when he finds the pride he has taken in his trees to be changing to dismay as they rapidly increase in size year after year, and shut out desirable outlooks and air, till at last they become a nuisance instead of a joy.

For all purposes where a dwarf tree is needed *K. paniculata* can be recommended and without reservation. It is easily grown in common soil, and is believed to succeed best where the ground is not very rich. It has irregular, spreading branches covered with dark green leaves alternate and deeply toothed. In June and July, as the blossoms of the horse-chestnut fall and flowering trees become rare, those of koelreuteria appear. They are mostly of a rich yellow, and are borne in large panicles of many flowers that stand out conspicuously from the green foliage. These are followed by a fruit varying in color, green, bronze-red, and purple. The keeping quality of the leaves, and the succeeding flowers and fruit, make the tree ornamental and interesting for an unusually long period. The fact that the young plants spring up readily from the seed that falls to the ground, leads to the hope that this beautiful little foreigner will some day be familiarly known in our gardens.

RHUS—Sumach.

THE genus rhus includes more than a hundred species of interesting plants widely distributed over both continents, though few or none are found within the tropics. They are most abundant in the United States, China, Japan, and South Africa, the range on the

Rhus—Sumach.

Dark Continent extending to the Cape of Good Hope while on the other hand it reaches to equally northern extremes. Many sorts are tender, while others are comparatively indifferent to heat and cold, as well as to soils and other horticultural conditions. A few are exceedingly poisonous, in this respect rivalling the famous upas tree, but nearly all are not only harmless but ornamental. These sumachs, as they are popularly called, are prized for their beauty of foliage and their peculiar flowers and fruitage; and when rightly handled they may fill an important niche in garden and park planting.

R. aromatica—fragrant sumach—is a low-growing shrub with a spreading head, and rising five to eight feet. The leaves are slightly pubescent in their early stages, thickening with age, and when crushed give out an agreeable odor. They are compound, with three pairs of leaflets unequally cut and toothed. The flowers are pale yellow, in spikes closely clustered, and appearing in advance of the foliage in April and May. Though an interesting species, it is not the most showy of its class in either leaf, flower, or fruit.

R. typhina, or, as it is commonly called, the staghorn sumach, is a much more ornamental shrub, and is also of American origin, being quite frequently found growing in the borders of woods and on dry and infertile hillsides. In fact, it does not appear to be very particular as to soils and surrounding conditions. The branches are blunt and clumsy-looking, usually thickly covered with hairs, and almost wholly lacking in symmetry of arrangement. For this reason the shrub is best planted among other sorts such as may partially hide the seeming deformity and yet permit the brilliant scarlet autumnal foliage to be seen. The leaflets, fifteen to thirty-one, are oblong-lanceolate, and very few plants at the season show to better advantage, or do more to enliven the border and brighten the landscape. The buds are also curious, as they are deeply set in the middle of a large leaf-scar, and protected by a mass of hairs, almost cone-shaped, against climatic exposures. The flowers are greenish yellow, and collected in a thyrsoid, terminal panicle. This is, to say the least, an interesting shrub, and can be used to advantage in many positions.

R. glabra, or smooth sumach, is one of the best known species in this country, being found along the borders of woods or growing freely in dry and sterile situations, often taking possession of entire fields and holding them with such tenacity, through its deeply extending roots, as not to be easily dislodged. It is a low-growing, spreading bush with irregular branches and a rather unshapely form, but is, nevertheless, of considerable horticultural value. The leaves are compound, often a foot or more long, with from fifteen to thirty-one leaflets on a large smooth stalk. The leaflets are nearly or quite sessile, oblong, and pointed at the apex, though rounded at the base. The blossoms are in large and much-branched heads at the ends of the stems, and of from ten to twelve inches extension. They are greenish-yellow and slightly fragrant. In the autumn the foliage becomes brilliantly crimson, and the heads of fruit, made up of velvety berries arranged in cones or spikes, are among the most showy productions of the field or forest. When growing in masses, especially if looked

Rhus—Sumach.

upon from a little distance, few plants are more showy. The fruit remains longer than the leaves, and though slightly changing its color is still attractive. *R. g. laciniata*, one of the varieties sometimes known as the fern-leaved sumach, is still more worthy of a place in the garden. It is a smaller plant, from three to five feet high, and has finely cut foliage. It is best grown in the form of a low bush with several stems which, if permitted, will spring from the common root in the form of suckers. As in autumn its feathery leaves take on the same crimson hues as the

RHUS GLABRA LACINIATA.

type, a cluster of these long stems when at the best, and in contrast with surrounding objects, has almost the appearance of a ball of fire. This comparatively new form is one of our most valuable recent introductions, and has been long enough before the public to have won a reputation accordingly. *R. copallina* is a dwarf sumach, with running roots, and is often scarcely more than a foot high. It can be made to quickly cover rocky and barren spaces whenever such a result is desired.

Ornamental Shrubs.

R. cotinus—Smoke tree, Venetian sumach—is not only the most remarkable member of the family, but also one of the most peculiar and interesting of the hardy plants. Its departures from the family type led Nuttall in his description of the bush to name it *Rhus cotinoides*, indicating thereby resemblance rather than identity. Prof. Sargent in our own day takes much the same view, and so inclines to a distinct classification. Be that as it may, the plant is likely to be known in the future, as it has been in the past, to the great majority of those who have to do with it, as the Venetian sumach or smoke tree. The bush is much planted in England, and the London *Garden* speaks of it from that standpoint as follows: "This when in flower always arrests the attention, because of its singular appearance, of even those who do not take a general interest in shrubs. On account of the feathery nature of the sterile flower-clusters, some call it the wig tree, a name by which perhaps it is better known than Venetian sumach. It is an invaluable shrub, as it is attractive at a time when shrubberies begin to look dull and monotonous. It is always a dwarf, spreading bush, rarely more than eight feet high. Its glaucous, round leaves make a pleasing contrast to the reddish, feathery clusters. It is hardy, almost evergreen, and grows in all kinds of soils, but must always have plenty of room to allow of full development." The autumnal foliage assumes a lovely shade of rosy-crimson, and whether in groups or standing as single specimens out in the open, it is certain to attract attention. The flowers are in loose panicles, of light purple or flesh color, the pedicels becoming

Rhus—Sumach.

lengthened and hairy after blossoming, and spreading over the whole bush, giving it a misty appearance, whence it is

RHUS COTINUS.

often known as the smoke tree, by which name in some sections many only know it. It is a native of Caucasus and other eastern countries, and proves hardy in all parts of the United States.

R. semialata—Osbeck's sumach. This, though introduced to eastern cultivation as a Japanese tree, is found to be widely distributed throughout most of the Asiatic countries, appearing frequently on the Himalayas and other lofty mountains as well as in lower altitudes. It grows in the form of a tree, twenty to thirty feet, and with somewhat remarkable vigor. Properly cared for, it makes a shapely and symmetrical specimen, and becomes an object of beauty wherever planted. As first described as seen at the Arnold Arboretum, where it had been tested for several years, it was represented as having compound leaves some fifteen inches long, with fifteen to twenty-three leaflets six or seven inches in length, dark green and shining on the upper surface but pale and covered with a soft, rusty pubescence beneath. The short-pediceled flowers are produced in large terminal panicles, the whole constituting a very showy head. The florets are pure white, though the projecting anthers give a yellowish cast to the cluster as a whole. Not much can be said for the fruit, which is inferior to that of several of the American species. The blossoms appear late in July, or in early August, and the fruit holds on until late in autumn. In Japan the coloring of the leaves near the close of the season is very vivid, and the most brilliant orange and crimson in all the forests. In the North the plant is not regarded as quite as hardy as some of the native species.

AMORPHA.

THIS is a genus of the order *Leguminosæ*, and a native of the United States, and, though confined by Nature's processes to the southern section, the cultivated species thrive as far north as New England, and are quite at home in the middle Northwest. They are described as handsome, hardy, deciduous shrubs, with graceful, pinnate leaves and many pairs of leaflets which are full of pellucid dots. The racemes of the flowers are in elongated spikes, usually in fascicles at the tops of the branches, and the corolla is without wings and keel. They are all well adapted to small shrubberies, preferring a sheltered situation and doing well in any good garden soil.

A fruticosa, or the shrubby species, is better known as the false or bastard indigo, from the color of its blossoms, which are a deep indigo-blue or very dark bluish-purple. The individual flowers are small, and, except on close study, appear dull and uninteresting; but closer inspection reveals the beauty of the richly colored petals as contrasted with the yellow anthers, which protrude slightly beyond the rim of the cup. When these are gathered into spikes borne at the terminals of the several branches, and these spikes are bunched in twos and threes, thus covering almost the whole bush, they appear to especial advantage, and are really beautiful. The bush itself is somewhat stocky, as it sends up numerous free-growing shoots to the height of six or eight feet, clothed with elliptic, oblong leaves, the lower ones on long petioles and the upper more nearly sessile, glabrous, and slightly pubescent. The plant is at its best late in June or early in July, when its blossoms are in perfection.

A. canescens—Lead plant.—This is a much smaller plant, seldom rising more than three feet. According to the older botanists, it was introduced to English gardens as long ago as 1812, and was received with considerable favor, and somewhat freely employed in decorative horticulture. But it is now seldom found in gardens on either continent, having been crowded out by the multitude of new and more rare, but not always better, sorts. Still it is named in several of the nurserymen's catalogues, especially on the European continent. It is worthy of a restoration to popular favor, and will doubtless again see better days. It takes its name from the color of the foliage, which in both stems and leaves is of a whitish or lead-colored cast. The flowers are much the same as in the larger species, and are also produced in terminal, clustered spikes several inches long. They are deep purple, and do not usually make their appearance until the first of August, thus affording color at a time when flowering shrubs are not supposed to be at their best. It is a native of Missouri, where it is still found growing wild.

Each of these amorphas has given off several varieties, which, however, do not appear to be especial improvements on the type.

OSMANTHUS.

THE name of this genus was applied because of the peculiar and agreeable fragrance of its blossoms, and is made up from *osme*, signifying perfume, and *anthos*, flower. It belongs to the olive family, and is sometimes classed under that head; but the best authorities point out differences that justify a distinct classification. The tribe is small, only seven species being named. Those of greatest value are natives of North America or Japan. Though classed as reasonably hardy, none of them will endure extreme northern winters without protection. But some of them, at least, will grow in the open, in the latitude of Philadelphia, and with slight care in New York and even in southern New England. They are certainly as hardy as the English holly or most of the mahonias, and should receive much the same treatment in cultivation.

O. americanus is a native of this country, and grows in tree form from five to six feet. The foliage is elliptic-lanceolate, the leaves about four inches long, thick and glossy. The blossoms are nearly sessile, in threes, axillary, appearing in June in a compact thyrse. These are followed by purple, globose fruit in the form of a nut, and quite palatable to the taste. It is a good plant, though not the best of the genus.

O. aquifolium.—This is a native of Japan, and esteemed one of the best. The foliage very much resembles that of the holly, being oblong or oval, coriaceous, smooth, and glossy. The leaves are stiffly armed with teeth, and are evergreen, thus affording an attractive winter aspect in connection with other broad-leaved evergreens. The flowers are white, very sweet-scented, appearing in autumn. Nicholson pronounces it a very handsome evergreen, varying considerably in the size and toothing of the leaves. *O. a. illicifolium* is a variety which is held to be an improvement over the type. It has smaller leaves and a more dense and compact head, and is supposed to be able to endure greater degrees of cold. It is pronounced by a good authority "the most beautiful of all the evergreen shrubs outside of the conifers." This is probably an overstatement resulting from undue partiality or exceptional conditions. But it is really a beautiful plant, and should be brought into much more general use. Its vigor is such that it may be safely planted in almost all sections of the country, numerous instances being reported where it has stood a zero climate without the loss or even turning of a leaf. Of this there are several forms of variegation in cultivation, mostly showing different colorings of leafage. *O. a. myrifolium* is what is known as "a fixed sport," with dense habit and leaves without spines. *O. fragrans* is perhaps the best known of all the species, though properly esteemed an olive and described as *Olea fragrans*. It possesses the general characteristics of the genus as already pointed out, its flowers being yellow instead of white as in the *O. a. illicifolium*. They appear in June and continue until August under usual conditions, and are very pretty as well as exceedingly sweet-scented. The foliage is good, somewhat lanceolate, and finely serrated, glossy above but paler on the under surface and less showy. It is a native of China, and grows to the height of eight or ten feet, making a very attractive bush or small tree, and worthy of a place in every garden where climatic conditions are favorable.

EUONYMUS.

THIS genus of plants, though not especially large, is widely distributed. It was known to the ancient Greeks, and is said to have received its name from Theophrastus by the combination of *eu*, signifying good, and *onoma*, name ; and it is thought that this term may have been applied because the plant had the reputation of being poisonous, especially to cattle and other animals which might feed upon its leaves. The application of a good name to an object as a cover for its bad qualities, was somewhat common in those days, perhaps for the charitable purpose of hiding defects and speaking well of the unfortunate ; or possibly that no offence should be given to the gods, whose workmanship it was not deemed prudent to criticise. However this may be, the euonymus is a good shrub, highly ornamental, and worthy of a place in all our gardens. That its bark, leaves, and flowers contain elements unfitting them for food is very true. Because of certain qualities the plant has long been esteemed by physicians for its medicinal properties, but these are not such as to render it dangerous to handle, or even to eat in small quantities. Dr. Gray mentions only two hardy species as indigenous to North America, and one variety.

E. americanus, or strawberry tree, is described as a low, upright, or straggling bush, from two to five feet high, with bright green ovate or oblong-lanceolate leaves, and greenish-purple flowers. The fruit, when ripened in autumn, is crimson-scarlet, and very showy. This variety is found mostly in moist situations, along the banks of streams, and is of service for shady situations as an under-shrub.

E. atropurpureus, known as the burning-bush, and also, in some localities, as the waahoo, is a shrub from six to fourteen feet high, with somewhat spreading branches, though sometimes grown in tree form. It is widely distributed throughout the country, but more sparingly on the Atlantic slope. The leaves are bright green, oblong, serrate, and nearly sessile. Flowers appear mostly in fours, are dark purple, somewhat showy, and come in June. But the chief attraction is the fruit, which remains long into autumn, and from its bright scarlet or strawberry color and great abundance comes the popular name of burning-bush. This is surpassed for late autumn effects by few, if any, plants in use.

E. europæus, the European spindle tree, is much the same, and is also known as burning-bush, though its seeds are of a paler hue, verging upon orange-yellow. This has long been a favorite in European gardens. It grows equally well here, and is largely planted. *E. latifolius*, the broad-leaved euonymus, is also a shrub of European origin, and grows six to eight feet high. The flowers are white when first opening, afterward becoming shaded with purple. They appear in June. The fruit is large, abundant, and highly colored. Though not much known in cultivation in this country, it is worth a trial in every good-sized collection. *E. nanus* is a veritable dwarf, growing about two feet, with spreading branches. It appears to be reasonably hardy, and answers a good purpose when plants of its proportions are desired. Its fruit is abundant and showy.

E. japonicus, Japan euonymus, is an evergreen, and

Euonymus.

not so hardy as the European and American species. Its leaves are oblong and sharply pointed, and the flowers, which appear in April, are small and somewhat fringed.

EUONYMUS JAPONICUS.

The numerous branches are pendulous, gracefully drooping on all sides, and affording a full, round head. There are several varieties in cultivation, in one of which the leaves are margined with white, another with yellow, and still others with less distinct markings. It grows to a height of twenty feet. Meehan says : " It is one of the few shrubs doing well along the seashore, though probably not hardy north of New Jersey."

E. radicans is also an introduction from Japan, and is coming to be much used in England and America. It is an evergreen climber of slow growth, more hardy than the English ivy, and for that reason better for some situations. Though somewhat particular as to soils and exposures, it possesses qualities which commend it when it is desired to cover small areas, and to establish edgings for paths and borders. When thus employed it can be kept as low as the box, and is to be preferred. The variety with variegated foliage, *E. r. argentea*, is especially desirable. The leaves are light gray, and hold their color throughout the winter.

E. yeddoensis is a native of Japan, and is a vigorous and compact grower, with large leaves deep green in summer and bright red in autumn. The fruit is scarlet, hanging from a pink envelope, and lasts until the foliage falls. It may not be entirely hardy in the Northern States, but should have a future in the South. *E. japonicus compactus* is scarcely known in this country. The London *Garden* describes it as very different from the commoner forms of Japanese euonymus, as it makes a neat, compact little bush not more than six inches high, but quite upright in growth. The oblong-shaped leaves are each about three quarters of an inch in length, of a deep shining green with a narrow margin of clear white. It will doubtless be sought after, when it becomes better known, as desirable for edgings and as a low bedding plant.

Ostrya—Hop Hornbeam, Ironwood.

E. alatus, or Japanese winged euonymus, proves to be large and well-shaped, with many of the best characteristics of the family, and in every way a most desirable garden plant. Its early buds in spring are marked with pink, which gives it a peculiar appearance. The foliage is agreeable at all times, in autumn turning to a pink or dark red hue. The flowers are white and not at all showy, but masses of brilliant scarlet berries afford an autumnal exhibit of rare beauty. It is not much planted as yet, but as it is quite hardy and easily grown it has only to be better known in order to win popular favor.

OSTRYA—Hop Hornbeam, Ironwood.

THIS is a genus of the order *Corylaceæ*, of but three known species. Until recently, there has been supposed to be but one in Europe, and one in America, but within a few years a second American form has been discovered in Colorado, though it is smaller and in almost every way inferior to those which have been longer and so much better known, and but little account in this connection need be made of its presence. It has been named *O. knowltonii* in honor of the discoverer, Mr. Frank H. Knowlton, and is interesting chiefly as a curiosity. The European species, *C. vulgaris*, has long been recognized in the Old World, and considerably planted, but now has practically given way to the American species, which answers a much better purpose.

O. virginica, or hop hornbeam, is a small tree, in the extreme North scarcely more than a shrub, of slender form, with foliage somewhat resembling that of the black birch.

The leaves are oval or egg-shaped, about three inches in length, tapering to a long point, and with many smaller ones on the same branch. They are smooth above, and slightly downy below, with somewhat hairy angles. The autumnal coloring is very fine, as the head assumes various shades of orange-brown or russet. The flowers appear with the leaves, and the fruit is in a closed, bladdery-like, oblong bag. These bag-like involucres form a sort of strobile, in size and appearance so like that of the ordinary hop cluster as to justify the use of the common name. It is a handsome tree, especially when in fruit. The wood is of the same general character as that of the carpinus, and is used for the same purposes; and like that it is sometimes called ironwood.

DESMODIUM.

THE desmodiums are shrubs comparatively little known, and yet they are easily grown in any good garden soil, and have the especial merit of blooming late in the autumn and continuously until cut down by frost. The branches are numerous, gracefully pendulous, and when covered with flowers the effect is very fine. As a rule, they should be cut back severely, and if every year to the very ground the roots will remain uninjured, and the following spring throw up vigorous shoots productive of the result described. There are two species suitable for cultivation in this country. *D. penduliflorum* has deep rosy-purple flowers which are very attractive, and by many it is placed among the best of all our ornamental shrubs. The fact that the flowers appear in

Syringa—Lilac.

September or early October, and continue until winter sets in, makes the plant a favorite wherever known. *D. japonicum*, a native of China and Japan, grows to about half the height, and produces white, pea-shaped blossoms the whole length of the dense, pendulous branches, and, like *penduliflorum*, late in September. It is an especially choice species, and worthy of introduction into all our gardens where autumn flowers are desired.

SYRINGA—Lilac.

THE genus syringa is too familiar to the general public to be in need of an introduction, though not many persons are familiar with its history or most of its numerous representatives. The species and varieties are almost everywhere known to the English-speaking people as lilacs, while the true name has been popularly applied to another and quite a different family of shrubs to which it in no wise belongs. Though a few sorts only have been widely distributed in cultivation, there are yet many others, including numerous later additions to the list, which are exceedingly valuable for horticultural purposes and are now coming into use.

The syringas belong to the natural order *Oleaceæ*, and as first known to botanists were supposed to be the peculiar product of Asia Minor and the countries bordering on and in the vicinity of the Black Sea. The first specimens were brought, as early as 1597, to England, where they were received with much favor and afterward largely distributed. From the mother country they very naturally found their way to America, and were among the first to be thus imported and used in the colonial gardens. Here, too, they were widely scattered throughout town and country, and soon became so common that many people have been led to suppose they were natives of the soil.

Not more than from twelve to fifteen distinct species are known, but these have thrown off a great number of varieties, some of which are far more interesting than the types from which they have sprung. As a rule, all the lilacs are hardy and vigorous-growing shrubs or small trees, and mostly with numerous long, straight branches and large clusters of various-colored flowers. Few shrubs or trees blossom more freely and uniformly from year to year, and fewer still adapt themselves to greater diversities of soil and climate. They are at their best in early spring when flowers are most welcome, and can scarcely be planted amiss wherever a flowering bush or tree of their proportions is desired. Though calling attention to a goodly number of forms, old and new, it may well be said that no list of the species and varieties can be complete and remain so for any considerable period, inasmuch as new hybrids and fresh offshoots are making their appearance almost every day, not all improvements on the older forms.

S. vulgaris is the old-fashioned lilac, which is understood to have come from Asia through the medium already designated. Until within a few years it has been supposed that it was nowhere a native of Europe, but of late, in some of its varietal forms it has been found growing wild in a Hungarian forest, though it is difficult to de-

Syringa—Lilac.

termine whether these growths are indigenous to that section or came from seeds or plants that had "escaped from civilization," and thus found a new home through processes well understood to horticulturists. The tree grows in somewhat bushy form from ten to twelve feet high under favorable conditions, with smooth, cordate leaves on long petioles, and large, showy panicles of flowers of that peculiar and almost indescribable color that takes the popular name of the genus—lilac.

SYRINGA VULGARIS.

As already indicated, scarcely any plant has been naturalized in so many lands, and been everywhere so thoroughly welcomed, as this, and it is not too much to say that it merits all the honors which it has received. It may now be old-fashioned, but it is nevertheless to be loved and honored.

Of this species there are many varieties, and among them some of the finest plants in cultivation. It would be practically impossible to describe them all or even name them in this connection, but the following are among the most desirable sorts, and the descriptions will be sufficient to furnish the basis for a wide selection. *S. v. alba* is the well-known common white lilac which so many have learned to love, as it is already widely distributed. It closely resembles the type in everything except the color of the blossom, which is pure white and in large trusses. When the two are growing together the contrast in this respect is very marked. Most, if not all the other, white lilacs are of inferior proportions to the *vulgaris*, while this is at least of equal height. *S. v. alba grandiflora* is a much smaller bush, five to six feet high, but has larger clusters of even more showy flowers. Another white variety, *S. v. Marie Legraye* is a veritable dwarf, three to four feet, but though low is sufficiently bushy to show a beautiful head of pearly blossoms, making it especially desirable for borders or single specimens where there is but little room to be devoted to lilac culture. It is to many a curiosity greatly admired. *S. v. Frau Dammann*, another offshoot, is pronounced by some good judges the very best of all the white sorts, but, though possessing many excellences, it is scarcely entitled to that distinction. It would be difficult, indeed, to name any member of the family as to be preferred over all others without regard to conditions and the effects to be desired. This is, however, a choice form, as the truss is very large and the color without spot or blemish.

But, though the white-flowering varieties are best known and most largely grown of any of the departures from the typical color, there are other shades equally beau-

Syringa—Lilac. 215

tiful and useful in garden planting. There are those, too, of different times of blossoming, some earlier and others

SYRINGA, LUDWIG SPAETH.

later, thus giving such as desire to extend the lilac season an opportunity to do so by at least a month's prolongation. Among these, attention may first be called to the variety

Charles X., which, so far as known, is a favorite with all planters. It grows from six to eight feet high and in good form. Under favorable conditions it is a vigorous grower with glossy foliage and large open trusses of reddish-purple blossoms. This sort has gained a popular favor as one of the most excellent of its class, a reputation which it well merits. Among others of about the same height, differing chiefly in the color of the blossom, may be named *Ludwig Spaeth*, dark reddish purple; *gloire de moulins*, purple-rose; *Comte Horace de Choiseul*, deep violet-red; *Prof. Sargent*, red and violet; *pyramidalis*, azure-rose; *rubra insignis*, rosy purple; *Senateur Volland*, rosy red; *Jean Bart*, rosy crimson; *Van Houttei*, red; *Jacques Calot*, rosy pink; and *cærulea superba*, blue. This list might be almost indefinitely extended, though in most cases the difference would be found so slight that it might tend to confusion rather than profit.

S. persica—Persian lilac.—This was probably the second of the lilacs brought to America and accepted as one of the favorite plants for popular planting. It is a much smaller bush than *S. vulgaris*, usually growing but four to six feet high, with numerous slender branches, all of which are of service in the production of flowers. The leaves are small, lanceolate, holding their bright green color through the entire summer, so that the shrub is always attractive. The flowers are purple with a bluish cast, and very pretty. They appear in April and May, according to local conditions, and always make a fine show in the border or hedge as well as in groups or as single specimens. *S. p. alba* is a variety with white blossoms,

Syringa—Lilac. 217

and differing from the original in little else. *S. p. laciniata* is a form with finely cut foliage, but, though interesting, is not especially to be preferred. The Persian lilacs are better suited to forcing under glass, for the production of cut flowers, than most others, and, thus grown, the blossoms are great favorites for the Christmas and Easter holidays.

S. chinensis — Chinese lilac.—The nativity of this

SYRINGA CHINENSIS.

species, or as some call it, variety, is not certainly known, though it is believed to have come from China, and thought to be a possible hybrid between *S. vulgaris* and *S. persica*. So far as known, it is found in gardens only, and as a cultivated plant. In proportions it is intermediary between the two species named, and is found possessing some of the special characteristics of each. It grows to a height of seven to eight feet, with a loose, open head. The leaf is thick, ovate, and in early spring inclined to be glossy. The flowers appear in large open panicles and are of a reddish-purple or violet color somewhat peculiar to the family. It is also known in the catalogues as *S. rothomagensis*, though the identity of the two is not conceded by some authorities. However that may be, *Garden and Forest*, in one of its earliest issues, fixes its place in practical horticulture in pronouncing it "one of the hardiest and handsomest shrubs in cultivation, producing its enormous clusters of flowers in the greatest profusion."

S. oblata is also of Chinese origin, and is sometimes advertised as a new species in cultivation, but it is not such in any proper sense of the word. And yet, though not new, it is quite rare even in our best gardens. It was discovered by Mr. Fortune in a garden at Shanghai, and through him introduced many years ago to Europe, where it was welcomed as one of the best of its class in cultivation. It is nowhere found growing wild, and so must be considered strictly a garden variety with an unknown parentage. In its general appearance it somewhat resembles *S. vulgaris*, having broadly cordate and sharply pointed leaves, which hold on until late autumn, and often assume brilliant shades of color, constituting it at that season one of the most attractive objects in field or garden. The flowers are large and of a purple shade, somewhat differing from that of the common varieties, but not easily described. They appear ten or twelve days earlier than those of most of the species, before or with the expanding foliage, thus serving a good purpose in lengthening the season at the

Syringa—Lilac.

beginning. The species is reputed hardy and capable of doing good service in warmer latitudes than most others.

S. villosa is another Chinese sort, having first been seen by a French missionary, about the middle of the last century, in the vicinity of Peking, and by him sent to his own country. This is also sometimes advertised as a new species, but it is so only in the sense of being rare in the section where it is sought to be distributed at that particular time. It has been known to horticulturists and to Old World gardens for more than a hundred years, and is another of the dwarf varieties, growing from four to six feet, and in a somewhat bushy shape. The plant carries large panicles of flowers which are light purple in the bud and when opening to the sunlight, but when fully expanded they become a creamy white. One of its excellences is found in the fact that it blossoms late in May, fully two weeks after most other sorts. The foliage is especially good, much resembling that of the well-known white fringe, thick, leathery, and glossy.

S. japonica — Japan lilac.—This is one of the later introductions from that land of flowers which has done so much for our modern horticulture. It is one of the largest of all the lilacs, at times growing with but a single stem some twenty feet high, and producing a spreading and well formed head. Because of this peculiarity it has come to be popularly known as the tree lilac, a designation not at all misapplied. The leaves are large, ovate, sharp-pointed, smooth above, and slightly pubescent beneath. The flower clusters are also of unusual size, in immense trusses of pure or slightly creamy white, but without fragrance. They do not break forth until near the first of July, thus adding from two to four weeks to the season when these beautiful plants are in bloom. Established singly on the lawn or in the field, this, all things considered, is the most striking and showy of the whole family. It is of fairly rapid growth, but should be given a good, rich soil to make sure of best results. So far as observation in this country has gone, it steadily improves with age, thus affording promise of long-continued usefulness as one of our best large ornamental shrubs.

S. josikæa.—The discovery of this species was made by the Baroness Josika on the banks of a rocky river in Transylvania, and so it took that lady's name. It was at one time supposed to have been a mere garden variety, but later information is to the effect that the home of the species is in the depths of a Hungarian forest far away from the haunts of man, where it is said that "for miles and miles where *S. josikæa* grows neither a house nor a hut is to be seen." And even there it was not freely distributed, and must have been of comparatively recent origin or else has been exceedingly non-aggressive in its character. At its best, it grows about five to seven feet high, and has bluish-purple flowers in comparatively small panicles, blossoming in May among the earliest of its class. The foliage is large, slightly wrinkled, very bright green, and almost glossy on the upper surface while quite light on the under side. The botanists are not quite agreed whether this is an independent species or a slightly varying form of *S. emodi*, a member of the family longer known, and which came from the Emodus moun-

Syringa—Lilac.

tain of the Himalayan range, from which it takes its name. It is thought, also, to be confounded with *S. villosa*, and for equally good reasons. These may be important questions for the consideration of the scientists, but to the practical horticulturist it is enough to know that the three so closely resemble each other as to size, habits of growth, and similitude of flower, that there is little or no choice between them. Each is a good garden plant and especially adapted to small grounds where larger growths might prove out of place.

Within a few years, a new race of lilacs has been brought to the attention of the public through the well-directed efforts of certain botanists and horticulturists, especially of M. Lemoine of Nancy, France, whose name had already become well known as a scientist of no mean repute. This gentleman, having entered upon a thorough process of crossing and hybridization in this direction, has succeeded in bringing out a class of these shrubs which, while retaining most of the better features of the genus, are still a great improvement upon the original forms. These have been given to the public, and have added much to the interest and pleasure of lilac cultivation throughout the world. Already the series of hybrids and crosses have been widely distributed, and specimens are to be found in many of our best gardens. As this method of production is continued from year to year, other new forms are constantly appearing, though most of the later sorts so nearly resemble some of those previously sent out that the limit has been nearly reached. The following are among the best of the recent productions having double flowers: what is known as *S. lemoinei*, named in honor of the originator, is not only one of the earliest of these new creations but also one of the best. It grows six or seven feet, with foliage resembling that of *S. vulgaris*, and has clusters of flowers eight or nine inches long, of reddish-purple color, distinctly shaded. The individual blossoms constituting the thyrse are double, quite large, fragrant, and so compacted as to show to the best advantage. That known as *President Grévy* has still larger panicles, measuring from ten to twelve inches, which are made up of individual flowers very double, three quarters of an inch across, purple, and with perhaps the deepest shade of blue extending over the whole that is to be seen in the lilac family. *Pyramidalis*, as its name indicates, appears in upright pyramidal form, with an abundance of rose-tinted purple blossoms which are carmine in the bud, and especially valuable for forming contrasts. *Lamarck* is an excellent variety with very large clusters, rosy lilac, and one of the most valuable of the entire group. *Michael Buchner* is a veritable dwarf of but three to four feet, but produces large panicles, which stand erect on the bush and are very showy in color, pale lilac.

Marie Legrange is one of the best dwarf varieties, and is well suited to growing in pots and to moderate forcing under glass. It grows freely, sending out numerous slender branches, all of which are crowded with blooms, giving it somewhat the appearance of a huge bouquet. These flowers are of the purest white, of good size, and borne in large trusses. Such a plant makes a fine show in mid-

SYRINGA ROTHOMAGENSIS.

winter, at Christmas, or at Easter, and can easily be in its perfection at either date. *Leon Simon* is of all the numerous double-flowering sorts one of the very best. The flower clusters are compact, with petals slightly incurved, and of a light purple shade, quite distinct in that respect. The individual blooms are of somewhat unusual proportions, and, gathered in the truss, constitute a long and characteristic bunch of the fairest proportions. This variety grows to a medium size and blooms in the height of the lilac season. *Alphonse Lavalle* is another of the dwarf forms, with double flowers of the old-fashioned lilac color, and in large and dense spikes. It, too, makes an excellent pot plant, and is well suited to forcing.

BACCHARIS—Halimifolia.

THE baccharis, sometimes known as the groundsel tree, is especially adapted for planting at the water's edge. It is grown freely in salt marshes, and is valuable to planters by the seashore; and as we have so few plants that will endure in such locations, it is often especially useful. It is a tall, resinous or glutinous shrub, growing sometimes to a height of ten or twelve feet, but usually smaller. It has dark green, abundant foliage, and small heads of white or yellow blossoms. These are not large, or very showy, but they are produced in such leafy panicles at the end of the branches that the plant makes a good appearance late in the season. This is especially the case with the fertile one, for in baccharis the male and female flowers are borne on distinct plants. The tufts of long, pure white hairs give to the

Andromeda. 225

females an entirely different appearance from that of the yellow-flowered males. The baccharis is seldom seen except at or near sea-beaches, but when once known in garden cultivation it will no doubt become more common.

FRINGE TREE (CHIONANTHUS VIRGINICA). See page 239.

ANDROMEDA.

TO the popular mind the name andromeda covers a large number of choice plants now distributed through a wider classification. According to most of the trade catalogues, it has been made to include a group of small shrubs which, though closely related, are now assigned to other genera, and given a distinct place in botanical lore. For sake of accuracy the newer and stricter classification will here be recognized and applied, while keeping in mind the popular conception sufficiently to prevent confusion on the part of such as have been familiar with the discarded name, and might be at a loss to recognize old favorites under the rearrangement of titles. No more interesting and useful plants for ornamental purposes than those here described can be found, and they will prove as meritorious under the new titles as under the old.

The name andromeda was first applied by Linnæus to a small semi-aquatic plant of the order *Ericaceæ*, discovered on one of his exploring tours in the North, and the great naturalist was seldom more enthusiastic in his praise of plant or flower than when he wrote of the water andromeda, and described it in his *Tour of Lapland*. " The flowers are quite blood-red before they expand, but when fully grown the corolla is of flesh color. Scarcely any painter's art can so happily imitate the beauty of a fine female complexion ; still less could any artificial color on the face itself bear comparison with the lovely blossom. As I looked upon it I could not help thinking of Andromeda as described by the poets, and the more I meditated on their descriptions the more applicable they seemed to the little plant before me, so that if these writers had it in view they could scarcely have contrived a more apposite fable. This plant is always fixed in some turfy hillock in the midst of swamps, as Andromeda herself was chained to a rock in the sea which bathed her feet as the fresh water does the root of this plant."

A. polifolia, the species so poetically described by the flower-lover, is indigenous to America as well as to northern Europe, where it is often called wild rosemary. It is found in wet, boggy land alike in New Jersey and Minnesota, and even farther north. It is not often that specimens of more than twelve to fifteen inches in height are seen, and on drier land they are even less high. The foliage is composed of linear, sharp-pointed leaves, entire, and with somewhat revolute edges. The midrib is prominent, and the veins reticulated. The flowers are mostly white, and tinted with pink and sometimes tipped with red. There are several varieties, differing chiefly in the color of the blossoms, but all possessing the same general characteristics.

The wild rosemary is the true andromeda. The pieris, cassandra, zenobia, and leucothoë, members of the popularly called "andromeda group," are now described under their own heads.

PIERIS.

ALL the members of the "andromeda family" bear classic names, and this title was conferred in honor of Pieria, the town in Thessaly where the Muses were supposed to congregate and pass much of their time. Not more than ten or twelve species are included in the classification, and it is not easy to determine the number known to the literature of mythology. The pieris is now distributed over portions of China, the Malay Islands, Japan, and North America, and includes some of the most interesting plants known to horticulture. Nearly all are hardy, and while especially adapted to cultivation in the middle sections of the United States, they can be grown in New England and many parts of the Northwest.

P. mariana, the stagger bush, is a small species from two to four feet high, and is capable of good service in enlivening the border or brightening the lawn in the early season when so few hardy shrubs are in flower. It is a native American, with mostly glabrous, oval leaves two to

PIERIS MARIANA.

three inches long, and graceful, nodding, white flowers borne on terminal, naked stems and branches in April and early May. It grows wild in wet or low grounds from New England to Florida, and westward to Texas and Arkansas, and may be considered hardy throughout the whole country. Its foliage is believed by the farmers to

Pieris.

possess poisonous qualities that are fatal to lambs and young calves, but it is doubtful if this is the case. *P. ligustrina* is another of the early bloomers, and grows five to ten feet high, and with a well-proportioned and compact head. The foliage is oblong-lanceolate, somewhat pubescent, and in panicles. The flowers are in small but numerous bunches, and cover the bush slightly in advance of the others. There is a southern variety still more pubescent, but it is not known to be hardy in the Northern States, or to be superior to the original.

P. floribunda.—No one should think of planting any considerable number of ornamental shrubs without including this charming plant in the list. It belongs to the class whose foliage continues throughout the winter, and is handsome all the year. It grows from four to six feet in height, and nearly always in a well-rounded and compact form. The branches and branchlets ordinarily spring from the ground, and, being numerous, the foliage is so dense that they are almost concealed from the eye. The flower buds, which are formed the previous autumn, occupy a prominent position above the evergreen foliage, and are largely proof against wind and cold. Like most others of this family, the flowers are white, and closely resemble the lily of the valley, hanging in pendent and much-branched racemes. While very desirable for out-of-door cultivation, this variety is also recommended for growing under glass when the flowers are wanted out of season, especially for Christmas and Easter holidays.

P. japonica, though a native of the distant East, is a near relative of the preceding, and with numerous close affinities. It, too, has rich glossy leaves which remain through the winter. They are from one to two inches long, and taper to a point at both ends, and are of such a striking character as to attract attention at all seasons. In the spring the new leaves are bright pink or red, and when seen at a little distance make the bush present the

PIERIS FLORIBUNDA.

appearance of one already in flower. This tint gradually changes to a light green, and later to the normal color which is much darker, and continues all the year. The waxy-white flowers are in long racemes, and borne in profusion so that in April or early in May the upper part of the plant is almost completely covered. In some respects this species is to be preferred to the *floribunda*, but a

Cassandra.

group of either planted in a bed or border edged with lower forms, such as heaths and azaleas, can be scarcely excelled in beauty. There is a variegated form with leaves deeply and irregularly marked with creamy-white. This combination of colors with the bright red of the early-growing leaves, and the glossy green of the mature foliage from the previous season, taken with the sheets of white blossoms, make some of the most remarkable groupings to be found in our parks and gardens. *P. formosa* is a Himalayan plant resembling the Japanese very closely. It is a beautiful bush clothed with broadly-lanceolate, evergreen leaves of a deep tint. The flowers are porcelain white and disposed in terminal, branching clusters. The buds of the pieris practically mature the previous year, and hold themselves in readiness to proceed with the utmost promptitude on the approach of spring to improve their opportunities. But were there no flowers it would still be one of the best foliage plants that we have, even surpassing the rhododendron in the bright glossiness of its leaves and in its general appearance. It is probably hardier than the rhododendron.

CASSANDRA.

THE cassandras, also popularly classed as andromedas, constitute a small genus of plants desirable in cultivation, and steadily coming into more general use. The name is in memory of the fabled daughter of Priam and Hecuba. There is probably but one species, *C. calyculata*, and but one or two varieties calling for attention in this direction. The typical plant is a native of the United States, ranging from the Carolinas to New England, though ordinarily not very freely distributed. It is everywhere a low and much-branched shrub with elliptic-oblong foliage, smooth above, but of a rusty brown beneath, thick and almost leathery, retaining its vitality late in the season, becoming in the South almost an evergreen. The flowers, which come out in April or May, are pure white, on short pedicels, oblong-cylindrical, and always attractive. They mostly spring from the axils of the leaves and are borne in one-sided racemes that cover almost the entire bush. It grows but from one to three feet high, and for best effects should be planted in groups or in connection with other shrubs.

C. angustifolia is sometimes referred to as a distinct species, but may be more properly described as a variety which is in some respects an improvement on the original form. It has longer and more narrow foliage, and slightly varies from the type in the shape and disposition of the blossoms. These last are in recurved racemes of the purest white, with an oblong-ovate corolla slightly contracted at the mouth. They, too, are on short flower stems, and borne at the ends of the branches, appearing in April. *C. crispa* is one of its synonyms, though sometimes catalogued as a distinct form.

ZENOBIA.

THIS is a monotypic genus of the *Ericaceæ* and, though a beautiful plant, is not so widely distributed as its merits would justify. It was named after the famous Empress of Palmyra who lived in the third century, and in her honor, having been even then brought into cultivation and won distinction. It is a small, well-shaped shrub of from three to five feet in height, and sufficiently hardy to endure our ordinary winters. It prefers a mixture of sand and peat, and in general cultivation should receive much the same treatment as the rhododendron or hardy azalea.

Z. speciosa.—This is a dense-growing shrub with foliage of pale green, slightly rounded, and holding its color well through the season. The flowers are small, bell-shaped, pure white, and of wax-like texture, borne in profusion in midsummer, and covering the entire bush. The shrub continues in bloom a long time, and is always pretty. As growing among the heaths, azaleas, rhododendrons, and other small plants it appears to especial advantage, and serves a useful purpose in lengthening out the flowering season. *Z. pulverulenta,* or *Andromeda dealbata,* is a variety that is prized by many even above the original. It, too, flowers long in succession, and in July and August, when most plants of the "andromeda group" are out of bloom. It requires much the same treatment in the garden as does the azalea, but is even more in need of water during the dry season if the best results are to be had the following summer.

LEUCOTHOË.

THIS genus of *Ericaceæ* is named after the sea goddess, and contains about eight species which are closely related to the andromeda. The leucothoës are all small, hardy shrubs indigenous to North America, and preferring moist, peaty soils and a temperate climate.

They are deciduous, with alternate, oblong, or lanceolate leaves holding on late in autumn, and in warm climates well into the winter, becoming almost, if not quite, evergreen. The flowers are generally white, though occasionally tinted with pink or rose. They are desirable plants for masses, in borders, or as single specimens.

L. racemosa is one of the best-known of the group, and is probably the largest, and, all things considered, one of the most desirable. It is a native of the Southern States, growing as far south as Florida and Texas, where it produces elegant white flowers in April and May. They are in long racemes, waxy in appearance, and very fragrant. The bark on the young branches, which are mostly erect, is bright red, and in marked contrast to the rich green and glossy, oval, lance-shaped foliage. The bush grows from four to ten feet in height, and is hardy in the North, though the flowers are sometimes injured where late frosts are common.

L. catesbæi is a plant from two to four feet in height, with ovate-lanceolate leaves, serrulate, and borne on long, slender petioles. Its flowers are pure white and beautiful, but with what is to most people a rather unpleasant odor, which is slightly offensive when one comes in contact with the shrub in blossom, and sometimes when in its immediate vicinity. But at a little distance this is not perceptible, and its general appearance is so good that this drawback should not altogether preclude its cultivation. *L. acuminata* is a species with snow-white blossoms in great profusion. They are in very short but numerous axillary racemes, the corolla being cylindrical, ovate, and drooping.

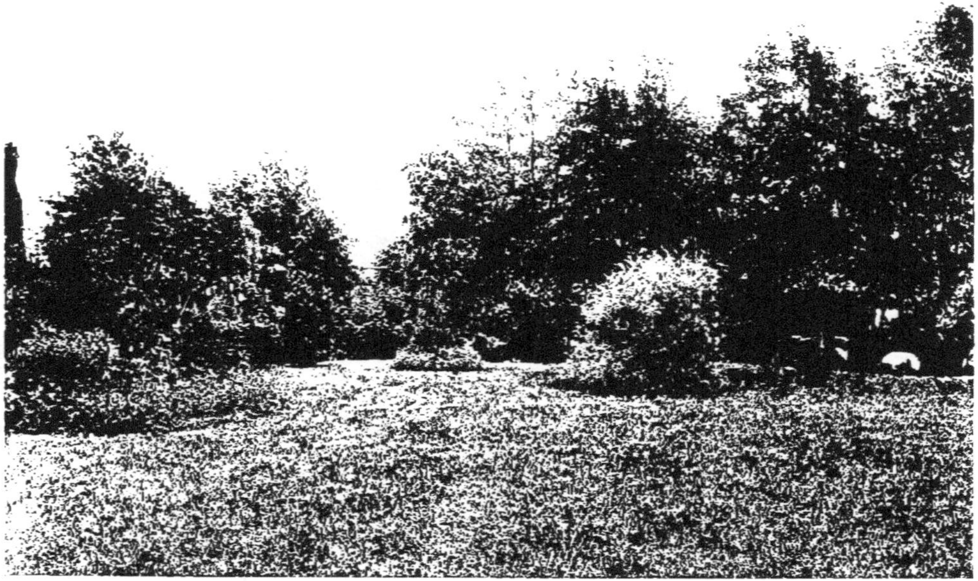

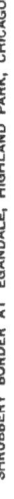

235 SHRUBBERY BORDER AT EGANDALE, HIGHLAND PARK, CHICAGO.

The leaves are ovate-lanceolate, gradually narrowing at the top, entire, serrulate, glabrous, coriaceous, and glossy. This is adapted to general cultivation in all parts of the country. *L. axillaris* takes its name from the fact that the flowers spring from the axils of the upper leaves or small branches, though this is scarcely a peculiarity of the family. They are white and much admired, especially as they appear very early, with their ovate, cylindrical corolla beset with scaly bracts. The foliage is oval, pointed, and marked with glandular hairs beneath, while the young branches are clothed with powdery down. The plant rises from two to three feet in height and usually blooms in May. *L. davisiæ* is a California species not much known in the Atlantic States. It is said to be one of the best of the class, having the usual recurved, white, pendulous flowers, and good foliage. Its hardiness has scarcely been sufficiently tested to justify promiscuous planting in the extreme north of our Atlantic slope. *L. recurva* is also a dwarf species, with little to especially recommend it unless a small plant of the kind is wanted for borders and rockwork. The flowers are good, of the typical color, and the branches more spreading than in most of the other forms.

ALNUS—Alder.

THE alders, a genus of *Betulaceæ*, grow chiefly in moist places and along the banks of brooks and streams, but all the members of the family can be transplanted without difficulty, and maintained in almost any reasonably good soil. The family embraces about a dozen species, six of which are natives of North Amer-

Alnus—Alder.

ica. These appear in the middle and northern portions of the United States and Canada, but seldom or never within or near the tropics. They constitute an interesting group of small trees or tall shrubs, though attracting comparatively little attention. When growing along the edges of streams they often do a valuable service in protecting the banks from washing by floods or being carried away by constant attrition. They are said to be largely planted for this purpose on the dykes and banks of canals in Holland and other low countries where such protection is needed. They are all the better adapted to this use from the fact that the roots do not extend far, but grow in a compact and knotted mass, throwing up a multitude of suckers near the original stem.

The wood is orange-yellow, soft, but exceedingly durable under water, and in Europe, whenever of sufficient size, is in demand for spikes and posts. It furnishes excellent charcoal for the manufacture of gunpowder, for which it has long been employed. The bark is in use for tanning and dyeing, and produces a reddish color if used alone, and with the addition of copperas, a jet black. The flowers are in terminal, imbricated clusters, the male and female in separate catkins on the same branch; the fertile ones being hard and compact, while the others are loose and open, both hanging long on the branches. The leaves are ovate, slightly lobed, with a blunt point at the extremity, and are smooth and somewhat glossy, often with white dots or scales. The flowers appear early, the ament having been formed in autumn, and so far advanced as to be in readiness for the first warm breath of spring.

A. glutinosa, commonly known as the European or black alder, is a native of northern Europe, Asia, and America. It is one of the most widely known as well as the largest and best of the family, often becoming a medium-sized tree. It has somewhat wedge-shaped leaves lobed at the margin, and almost black bark, especially when mature. The wood is of an orange color, and was formerly esteemed for the production of charcoal, and for use in the manufacture of small articles, as it is fine-grained and very hard. An authority of fifty years ago wrote: "It is one of the ornaments of many of the most exquisite landscapes of England. The dark green of its foliage, and the still darker hue of its bark contrast beautifully with the colors of other trees with which it is usually associated on the banks of our rivers." Within a few years a variety known as the cut-leaved alder, *A. g. imperialis laciniata*, has come into quite general cultivation, and is of deserved popularity among all who have become acquainted with its excellences. Like its type it makes a finely shaped, pyramidal tree with compact foliage and of somewhat rapid growth. The leaves are finely, not to say curiously, divided. It will thrive almost equally well in low, damp soils or on the hillsides in dry land. The species known as *A. cordifolia* is another of the larger sorts. It is a native of Italy, has very dark green and glossy foliage, and is said to grow rapidly in dry soils, and to be one of the most interesting of ornamental trees.

The smooth alder, *A. serrulata*, is also valuable for grouping, for, though it is seldom beautiful by itself, it helps wonderfully by its rich coloring in producing the

Chionanthus—Fringe Tree.

best effects on the lawn when placed in proper positions and intermingled with other varieties. In the South, it sometimes makes a tree thirty feet high, but farther north it is usually much smaller, and has a tendency to grow in clumps or thickets.

The speckled alder, *A. incana*, grows freely in New England. The younger shoots are brown and downy, and dotted with orange. As they advance in age, the bark turns to a bottle-green, and the dots become almost white. The leaves are large, oval, rounded at the base, much the same as those of the other kinds, but are slightly more serrate, being toothed at the termini of the principal veins. They are four to five inches in length, and three to four in width, standing on short, stubby footstalks.

The green or mountain alder, *A. viridis*, is a shrub of but three to eight feet in height. The leaves are round, ovate, sometimes downy on the under side, and sharply toothed. It prefers mountainous regions, and seldom suffers from the cold. Like most of the others it withstands high winds, and affords excellent protection for more tender sorts.

CHIONANTHUS—Fringe Tree.

THIS is a small genus of low, spreading trees, of the order *Oleaceæ*, which should rank among the hardy plants. The botanical name is derived from *chion*, the Greek for snow, and *anthos*, flower; and it is applied very fittingly because of the peculiar inflorescence which elsewhere has scarcely a parallel in nature. The best-known and most valuable species, *C. virginica*, is a native of Virginia and the Carolinas, and perhaps of other Southern States, but it proves hardy in cultivation as far north as New England and Minnesota, though in these extreme limits of cold it may always be well to give it as favorable a position as possible. The London *Garden* pronounces it perfectly hardy in England where it flowers with great freedom in May. It has also been introduced to the continental gardens and parks, and is noted as one of the curiosities of horticulture. "As an ornamental plant," says Professor Sargent, "the American fringe tree has much to recommend it; it is possessed of a vigorous constitution which enables it to flourish in regions of much more severe climates than that of its native home; its leaves are large, abundant, and excellent in color; it is not disfigured by insects or fungous diseases, and in May and June it is covered with long, drooping panicles of delicate flowers with elongated, narrow, nearly thread-like pure white petals."

The striking peculiarities of this shrub were observed and commented upon by the botanists who visited this continent during the early periods of its settlement, and found much that was new and at the same time worthy of admiration. Nor did it escape the notice of our

CHIONANTHUS VIRGINICA.

Chionanthus—Fringe Tree.

own people who were interested in the study of the forests and fields. As long ago as 1785 the distinguished Humphrey Marshall, a botanist as well as a statesman and jurist, wrote concerning it : " This shrub grows naturally in several places in North America, in a moist soil; rising to the height of fifteen to twenty feet, spreading into many branches, and covered with a light-colored bark. The leaves are large, oblong, and entire, placed nearly opposite. The flowers are produced towards the extremity of the shoots of the previous year, upon short, leafy, common footstalks, at the bottom of the leaves of which the proper footstalks come out ; and are divided for the most part into three parts, but often more ; each sustaining one small flower with four very long, narrow petals, which, when well grown, make a beautiful appearance. These are succeeded by oval berries of a livid blackish color when ripe, each containing one hard, oblong, pointed seed. The bark of the root of this shrub, bruised and applied to fresh wounds, is accounted by the natives a specific in healing them without suppuration."

About the only criticism that has been noted on this remarkable plant is based upon the fact that its leaves do not appear until so late in spring as to unfit it for artistic grouping with other sorts, since its branches are bare and apparently dead while everything about it is clothed with verdure. For this reason it may be objectionable where especial effects are sought for that particular period, and this should be considered in planting. But, though late in coming, the foliage when fully expanded is all that can be desired. The leaves are deep green and glossy, large and coriaceous, and it may well be doubted if, taking the whole summer through, anything will be found of superior excellence. And even the slight drawback will not apply when the tree appears in the plantation border or as a single specimen, for which it has peculiar adaptations.

It is not certain that more than one other species of the chionanthus exists, and that is a native of southern China and possibly of Japan, and is confessedly of less value than our native tree. *C. retusa*, for that is its name, is not only inferior in size, but has smaller leaves and shorter and less attractive panicles of flowers. It is in cultivation to some extent in Old World gardens, and is worthy of recognition, though not a worthy competitor on the score of merit.

LONICERA—Bush Honeysuckle.

THE loniceras are among the most useful of hardy ornamental plants known to cultivation. The genus includes a number of choice vines and climbers as well as shrubs, mostly natives of the north temperate zone — few or none being found within the tropics. But with this limitation southward, their range extends in the opposite direction well on towards the arctic circle. Some of the species are indigenous to Siberia, northern Russia, Labrador, and the region about Hudson's Bay in North America, and there is scarcely any place where it is desirable to plant them, however inhospitable, where they may not be expected to grow. They constitute a genus of *Caprifoliaceæ*, and the section known as bush honeysuckles, such as here described, are pretty well represented in our gardens, though some equally good varieties, and perhaps better, have not as yet been brought to the notice of the general public.

There are several American species, mostly small shrubs, which, though not so showy in fruit or flower as some of the foreign sorts, are worthy of more general recognition than they have yet received. Professor Gray, in his *Manual of Botany*, describes four of these bush honeysuckles as natives of the North, and capable of doing good service in many situations where especially hardy plants are called for. These are *L. ciliata*, a bush of from three to five feet high, which grows with spreading branches, oblong or heart-shaped leaves, and has greenish-yellow flowers in May. These are followed by red berries which continue long on the branches, and constitute it a really attractive plant for horticultural use in appropriate situations. It is a native of dense and rocky woods from New Brunswick to Manitoba and still farther northward. Another is *L. cærulea*, a smaller plant about two feet in height, having upright stems and branches, oval leaves downy when young but becoming glabrous later on, and producing yellow blossoms in great abundance, also in early spring. Its range is given as from Labrador to Minnesota, and southward to Rhode Island. In this case the fruit is purple and equally persistent. *L. involucrata* has oval or oblong foliage, yellow blossoms tinged with red, and dark purple fruit. It makes its home in the deep woods bordering Lake Superior and beyond. *L. oblongifolia* is a native of bogs and swamps of the same inhospitable region. Its leaves are two to five inches long, and sharp-pointed. The flowers are yellow and somewhat viscid, while the fruit is dark purple and enduring.

Of the imported species, the Tartarian honeysuckle, *L. tartarica*, is best known and most largely planted in this country. It is an erect, deciduous shrub usually five to six feet high, composed of a cluster of slender stems and branches which may increase in number with each season's growth. The foliage is oblong-cordate, of rather dull green, and in no way remarkable. The flowers appear in April and May, and are both abundant and beautiful, the yellow petals being somewhat thickened at the base, and rose-tinted. But it is the fruit that affords the chief attraction. This is ripened in July and August, and remains until late autumn. It is in the form of berries, in most cases nearly black, but in some bright cherry-red, and in others orange. These differences are not of a character to constitute distinct varieties, though doubtless by proper cultivation they might be extended and become fixed and permanent. These berries look as if they might tempt the palate, but they are found on trial to be inedible and even nauseous. It may be that for this reason the birds allow them to hang so long without interference on their part, and they remain to enliven the bush and give color to the garden until frost. There are several recognized varieties, differing chiefly in the color of the fruit, and not needing to be described separately. It is said that in Russia the horticulturists have been experimenting with them for many years, and have brought out some greatly improved forms, but they have not been given to

LONICERA MORROWII

the general public. The species, and, so far as known, all the varieties, are perfectly hardy.

L. fragrantissima has large white or pink blossoms nearly an inch across, which come out before the foliage is fully developed, and are exceedingly fragrant. It is a strong-growing and somewhat straggling bush of five to six feet. It came from China, and is one of the best for ordinary garden planting. The leaves are so persistent that it is sometimes classed as an evergreen, and in warmer climates its flowers are often borne in February and March. *L. standishii* differs but little from it except that the fruit is red.

L. hildebrandiana comes from Burma, and is the largest of known species, and altogether the most remarkable. It is an evergreen in the land of its nativity, but does not altogether prove such in less genial climates. General Collet, the discoverer, spoke of it as a conspicuous shrub with large, glossy leaves, and fine, crimson flowers seven inches long, and the experiments at the Kew Gardens appear to substantiate the claim of superiority. It is said to be much used in decorating the temples of its native country, and is looked upon there almost with veneration. It is not freely distributed here as yet, and can scarcely be found except in the hands of experts or in connection with public institutions.

L. morrowii is a vigorous shrub of from four to six feet in height, with spreading branches, and in July abundant yellow blossoms, followed by clusters of showy crimson fruit which is very persistent, making a fine show in the garden. One of its varieties, *L. frutea*, bears bright yellow berries, and is much showier than the original. *L. hispida* is somewhat pendulous, and has greenish-white flowers with purple fruit. It grows from two to three feet, and is a native of Siberia and south to central Asia. *L. mackii* reaches ten to fifteen feet in height, and is a much-spreading plant, making a large bush. The flowers are white, axillary, with a funnel-shaped tube. The leaves are ovate-elliptic, acuminate, rounded at the base, and of good color and substance. *L. xylosteum* is quite distinct in its foliage, grows about five feet high, and has pinkish-yellow blossoms in May before the leaves appear. They are cream-colored, fragrant, and very pretty. The fruit is crimson and of long continuance.

L. alberta, known popularly as the Siberian honeysuckle, is a dwarf about two feet high, and has special claims to attention where a plant of the size is desired. The slender branches have a drooping tendency, falling on every side, and are clothed with very narrow leaves two inches long and of a bluish shade. The blossoms, unlike most of the other species, appear in July and August, and, from a floral standpoint, the bush is at its best at a season of the year when flowers are scarce. They are also much larger than on most other honeysuckles, and deliciously fragrant. The color is a pinkish-purple and very effective.

PRUNUS—Flowering Cherry.

IN the popular mind, the cherries are ranked among fruit trees rather than among the ornamental trees. They belong to the order *Rosaceæ* and to the genus *prunus*, and by some have been assigned to a sub-genus, *cerasus*, under which they are generally catalogued by the nurserymen. The best authorities are now disposed to drop the latter name altogether, and, though the change may lead to some confusion for a time, it is so evidently in the line of correct definitions that it should be accepted by all interested.

The cherries are mostly natives of the north temperate zone, but are not altogether unknown in the tropics. They are mostly small trees, though some are large enough to be valuable for timber, and others are mere shrubs and bushes. Most of them are sufficiently hardy for ordinary horticultural purposes throughout the United States. In nearly all the species the flowers are white, single, and are borne in early spring. They are mostly in umbels, springing from scaly buds often in advance of the appearance of the foliage. In some of the varieties, especially in the more recently introduced Asiatic forms, the flowers are found to be double, and rose-colored or pink. It is largely through these varieties that the genus becomes valuable for ornamental planting. A few of these have been known for many years, but have not been sufficiently distributed to afford an opportunity to the general public to appreciate their value. All cherry blossoms are beautiful, and these later improvements are especially so.

One of the oldest species in use in this country is what is known as the bird cherry, *P. padus*, which is a native of Europe and Asia, growing to a height of twenty feet, of large spreading habit, and clothed in May with a great abundance of white, single flowers. These are followed by black berries ripening in autumn, and to some extent

Prunus—Flowering Cherry. 249

in use for domestic purposes. Whether in flower or fruit, this variety is always ornamental, and may be largely planted to advantage. It can also be grown in bush form as it often is where a smaller form is desired. *P. pennsylvanica* is the American wild red cherry, and more nearly resembles the European species than any other of our native sorts. It is especially noted for its reddish bark, and the red fruit which is very prominent in autumn. The flowers are white, and so numerous as to completely cover the tree, but do not continue very long in their perfection. It is distributed in Newfoundland, and the Hudson's Bay region, and throughout the Northwest, and was introduced to English gardens as early as 1773, where it is still in use as well as on the continent. *P. serotina* is another American species of about the same height, and is a well-known wild cherry, having white flowers in May and black fruit in August. *P. virginiana*, commonly known as the American choke-cherry, is very much like the European bird cherry in its general character, though not so tall a

DOUBLE-FLOWERING CHERRY.

grower. It makes a good show in the border. *P. avium alba plena* is one of the interesting varieties which differs from the established type in producing double flowers that are pure white, very large, and so numerous as to cover almost the entire branches in May before the foliage has become fully developed. It rises from twenty to twenty-five feet in height, with a somewhat spreading head, and in good form. *P. pumila pendula* is a unique form, and makes a twiggy growth, and when rightly cultivated constitutes an attractive shrub the season through. It grows from six to eight feet high, and the numerous white flowers are followed by bright red, acid fruit. For use to the best advantage it should be grafted on taller stems than are natural to it, when its branches fall gracefully but not to the ground.

P. sinensis flora plena.—This is one of the finest double-flowering cherries known. The tree is a native of China, and grows to a height of from twenty to twenty-five feet in good form, and is noted for its large white flowers, resembling miniature roses, which appear in great numbers along the stems and branches. This was the first of the Asiatic double-flowering species brought to this country, and so has been in cultivation here longer than the others. It proves reasonably hardy, though not an especially long-lived tree. *P. ranunculiflora* is another upright-growing cherry, having double white flowers which appear later in the season than most of the others of its class. It rises from fifteen to twenty feet.

But perhaps the finest ornamental cherries of all come from Japan. Some have already been introduced, though

Prunus—Flowering Cherry. 251

sparingly, in our gardens, where they command universal admiration. No people in the world take such interest in cherry culture as do the Japanese. The bursting of the cherry blossoms fills the souls of the people with delight, for of all flowers these are by them esteemed the most beautiful. They celebrate the occasion with great enthu-

CHINESE DOUBLE-FLOWERING CHERRY.

siasm, and give themselves up fully to the enjoyment of the beauties of nature. Professor C. C. Georgesen, a gentleman familiar with Japanese customs, and who had been present and participated in these festivities on more than one occasion, wrote some years ago in one of our magazines as follows:

"All classes of society, from the Emperor to the coolie,

rich and poor, old and young, all are enthusiastic admirers of the cherry flowers. The trees are planted in groups in the parks, in temple groves, in avenues lining many of the principal streets and roads, singly and in clusters in the gardens and yards about dwellings; they are dwarfed and grown in pots of all sizes, and trained in many forms and with pendulous branches; they are favorite objects with artists for conventional representation in paint, in lacquer, in metal—everywhere. Both in nature and art, one finds unmistakable evidence of the innate love which the people have for this flower. The trees bloom during the balmy month of April, when the raw and chilly winds of winter have given place to the warmth and calm of cheery spring, and all nature leaps into renewed life.

"Invited by the irresistible charms of nature, the people collect in gayly dressed throngs under the pink clouds of cherry blossoms, and there abandon themselves to jest and merrymaking. In Tokio, Ueno Park and the street called Mokojima are especially renowned for the charm of their cherry blossoms, and on pleasant days these places are visited by tens of thousands of people who have banished care and are bent solely on enjoyment, and they form, perhaps, the happiest collection of humanity that the world ever sees. It is a motley but always good-natured and orderly throng. The grave savant, the spectacled student, the flushed and prosperous merchant, the careworn poor, the decorous matron, giggling maidens and hilarious children—all are there, with laughing faces and in holiday attire. The cherry blossoms also usher in

JAPAN WEEPING CHERRY.

a series of private festivals which ministers of state, and the mighty in the land who glory in the possession of cherry groves, give to their friends. Even his Imperial Majesty, the Mikado, called by his subjects O'Tenshi, the Son of Heaven, is affected by the general impulse the blossoms impart, and issues a mandate to the effect that on a certain day, if it does not rain, he and the Empress will give a cherry-blossom festival in one of the imperial parks. Large, handsome cards inviting the guests are issued several days beforehand. The guests comprise all high officials of the government down to a certain rank, the corps diplomatique, foreign employees of the government at the capital who receive a salary of three hundred yen or more per month, high officers of the army and navy, and representative officers of foreign war vessels which happen to be in the Yokohama harbor. The writer had the honor of attending three of these parties, and can therefore speak from personal observation."

Of these Japanese forms the following are described as most interesting : *Prunus pseudo-cerasus* in its native country is said to form a large tree which grows wild in the forests of northern Japan and on the mountains of the south. It is described as somewhat resembling our sweet cherry trees in growth and appearance, but differs from them largely in its flowers and fruit. The wood is hard and fine-grained, and in general use for carvings and cabinet making. It has been a favorite ornamental tree with the Japanese from time immemorial, and through its long-continued cultivation a great number of flowering varieties have originated. Some of these newer forms are upright

Prunus—Flowering Cherry.

and stiff in habit, while others are spreading, short-jointed, and crabby, and still others have willowy shoots which lend themselves to various forms in which dwarf and pot-grown specimens are often seen. Professor Georgesen says that, as a rule, the earliest varieties in bloom are single, and the large and double flowers the latest to appear and remain the longest, though there are some notable exceptions to this. The flower and the leaf start at about the same time, but the leaf grows slowly at first, and the trees do not get green till about the time the flowers perish. These are mostly white and are quite large, appearing the latter part of April, and holding on well into May. Though there are numerous varieties in cultivation, all of them are charming and much admired by foreigners as well as natives. There has been some question as to how far north these plants may be cultivated with success, which is a question of much importance. Mr. J. G. Jack, having experimented with some of the varieties at the Arnold Arboretum, finds that they can be grown safely in the New England climate, at least in the region of Boston. But it is not to be doubted that they will succeed better in the southern Middle States and in the West. There is some question whether all varieties ascribed to this parentage are genuine offshoots.

Of the single-flowering varieties, what is known in Japan as the *kanzan* produces a plain white blossom which is very fragrant and, in that respect, an exception to the general rule. It is also an abundant bloomer, the flowers being very late and persistent. It has been grown in this country sufficiently to prove that it is adapted to our climate and conditions. The tree is bushy, spreading in habit, and has small leaves and slender twigs. There are several other single-flowering sorts, such as *nioi-sakura*, which has blush flowers, sweet-scented like the foregoing, and appearing later, thus helping out the cherry season. *Jishiu* is reckoned a single-flowered plant but has a tendency to become double. The flowers are blush, and spread wide open, showing numerous long red stamens with yellow anthers. It, too, is a profuse bloomer. *Koma-tome* has a very large, pure-white flower with broad petals and numerous long, showy stamens. The flower stems are also long, panicled, or branched. It is an early bloomer.

Of the double-flowering sorts, there is a long list from which it is almost impossible to make satisfactory selections because of their uniform merit. Among them is *fuzan-fukan*, the flowers of which are double, reddish-pink in color, almost globular in shape, and continue to bloom until late in May. *Botan-sakura* has reddish flowers, very large and double, at their best in the latter part of April. *Ko-fugen* is said to be the largest-flowering sort among the cherries. The blossoms often measure two inches or more in diameter, and are very double and very late, the color reddish-pink, growing lighter with age. *Kode-maru* is peculiar, as the flowers appear in very dense clusters at intervals upon the branches, and being short-stemmed they are crowded into balls which give the tree an unique appearance. In color they are light pink, the petals narrow and only moderately double. *Yo-kihi* has very large and very double flowers, color pink. The

Prunus—Flowering Almond.

flower stems are short, but the blossoms are scattered and do not form balls like the preceding. It continues to bloom till late in May. *Giyo-iko* is remarkable in that the flowers are of a clear, light green color when they expand, and gradually become tipped with a pinkish tinge. They are very large, and often panicled in long stems. The branches of the tree are slender and rambling. *Fugen-zo*, a popular variety, has very double rose-blush flowers on long stems, and is a prolific and long-continued bloomer. It is one of the handsomest kinds to be found in Japan, and is planted perhaps more generally than any other.

This list could be much extended, but these include the best varieties. " There can be no doubt," says Professor Georgesen, "that these cherries will do as well here in almost any part of America as they do in their native country, and as ornamental trees for the lawns and roadsides they will, while in bloom, surpass in beauty anything that we now have that blooms in early spring. Once before the public they will not lack appreciation."

PRUNUS—Flowering Almond.

P. amygdalus, popularly known as the flowering almond, usually appears in the nurserymen's catalogue under the head of amygdalus, but as it really constitutes only a sub-genus, or perhaps a group with the somewhat distinct peculiarities of prunus, it is well to conform to the true classification, and to count the almond as a species of that genus, as is now done by many of our best horticulturists and by nearly all the best botanists. The varieties will doubtless be known as almonds to the general public, and with good reason. The group is closely connected with the peaches, apricots, and nectarines, and in cultivation is subject to much the same conditions.

The common or wild almond grows from twenty to thirty feet high, and has light rose-colored blossoms, followed by fruit that is highly esteemed throughout the civilized world, and that calls for no description here. It has been in cultivation from time immemorial, and is yet found growing wild in northern Africa where almond groves, and even forests, still exist, especially in some of the Barbary States. From such as these the ornamental almonds and those of improved fruitage have doubtless sprung. While the type is not especially showy, the advance through cultivation is such that in 1892 the London *Garden* was led to say : " Of all the hardy, early flowering trees in the British Islands, perhaps the almonds are the most valuable from the point of view of ornamentation. In March and April no other tree produces such fine effects in the garden or park—at any rate in the southern counties of England." Since this was written several new varieties have appeared, superior to any then in cultivation.

P. japonica alba plena is one of the showiest sorts. It is of dwarf habit, and has an abundance of beautiful double white flowers, and is occasionally found in our northern gardens under the old name of flowering almond. It is capable of larger use in that section, and is very desirable farther south. Its companion plant, *P. japonica rubra plena*, is much the same except that its blossoms are red or rose-colored. They are very abun-

Prunus—Flowering Peach.

dant, and extend along the slender branches and twigs in advance of the foliage.

PRUNUS—Flowering Peach.

ACCORDING to tradition the peach, *Prunus persica*, came from Persia, and so it is credited to that country, though the place of its real origin is unsettled. It was formerly given the distinction of a

DOUBLE-FLOWERING PEACH.

separate genus, but is now held to be simply a species of *prunus* along with the almonds and plums. It is prized as one of the most valuable fruit-bearing trees under cultivation, and has thrown off a few double-flowering varieties with blossoms marked by a beauty and delicacy not often surpassed. That known as *P. persica alba plena* is very conspicuous, producing large double white flowers at the usual time of peach flowering. They appear in great profusion, and while they continue are always objects of admiration. This is heightened when the tree is planted in groups with others of the same character, such as *P. rosea plena*, with its double, rich rose-colored flowers; *P. sanguinea*, which is much the same except that the inflorescence is more a blood-red; and *P. versicolor plena*, having flowers red and white, variously marked on the same tree. Such a collection on the lawn or in the border, in front of other shrubs and trees, can scarcely be equalled in beauty and interest. There is also a charming purple-leaved variety of *P. persica* with foliage deep blood-red in spring, and becoming purple in summer.

P. davidiana is a comparatively new plant in general cultivation. It is of Chinese origin, though found later in Japan and some other eastern countries, and was first brought to the especial attention of the botanists by Abbé David, from whom it takes its name. The plant resembles the peach and apricot in its habit of growth, and may well be counted a member of the same branch of the prunus family. It is usually trained in bush form with slender branches, and it is on those of the previous season's growth that the flowers are produced. They come in very early spring, are nearly or quite sessile, and in the greatest possible profusion, pink and white, double, and slightly fragrant. When at the best this is a most attractive shrub or tree, and it is certain to become a favorite. It is reasonably hardy in the Northern States, though it is well to give it slight protection. The peach blossom is short-lived, as are most fruit blossoms, and the wise planter will have regard to this fact from the beginning.

PRUNUS—Flowering Plum.

P. maritima, known as the beach plum, is a somewhat common shrub along the Atlantic seacoast, and is worthy of more attention than it ordinarily receives. It is found in several varieties differing little and in unimportant particulars. As seen growing among the sand-heaps along the shores, it is often a mere straggling bush three to four feet high, and without form or comeliness, but when transferred to better soils, and given proper care, it not infrequently rises to a height of eight feet. Under cultivation it becomes a shapely and useful shrub, and is abundant in fruitage, which by some is highly esteemed. The stem is almost black, sometimes erect and sometimes prostrate, and with ash-colored dots. The branches are stiff, often dotted with orange, while the leaves are closely set, and are covered with a soft down. The blossoms come in advance of the foliage, and, though having no especial beauty, are interesting.

The chief virtue of this plant, however, is its adaptation to inhospitable situations, and its power of endurance where others fail. Whatever else may be said in its favor, it stands pre-eminent as a nurse plant in the peculiar positions where it is able to thrive. In shore planting one may begin almost at the water's edge, and follow with more desirable plants suited to such localities though in a less degree, and thus proceed until the most barren situations can be covered with herbage and made attractive. The beach plum answers for this good purpose in the vicinity of the inland lakes as well as near the salt water, and, indeed, in exposed and sandy situations everywhere.

P. pissardii is understood to have come from Persia, and is sometimes known as the purple-leaved Persian plum. It is a somewhat recent introduction to American gardens, but is already widely distributed, and universally popular, though its blossoms are quite inconspicuous, and of little ornamental worth. It is prized chiefly as a small tree or shrub, and has both colored bark and foliage. Perhaps in this respect it is without a rival in its class. In springtime the bark on the new growths is deep purple, and the bursting leaves as they come from the buds are violet-crimson. As they mature they take on a darker hue, equal to that of the finest of the purple beeches, and this they hold during the entire summer. Most trees and shrubs with such distinct foliage fade under the influence of the sun's hot rays, and become dingy, but such is not the case here. The flowers are small and single, and the fruit, which ripens in early autumn, is correspondingly inferior, and is scarcely visible as it is of the same color as the leaves.

P. triloba, a Chinese shrub with three-lobed leaves and somewhat spreading branches, only needs to be known in order to be admired. It proves well adapted alike to North and South, and thrives in a much wider climatic

PRUNUS MARITIMA. BEACH PLUM.

range than do most other varieties, and is not fastidious as to soils and situations. It grows from four to six feet in height, with numerous slender branches which in early spring, before the foliage appears, are covered from end to end with double, light pink blossoms about an inch across, completely covering the whole bush. These come in May as far north as New England, sometimes breaking out in March among the first harbingers of the summer.

P. watsonii is the sand plum so well known in some sections of the middle West, where it thrives in thickets on low, sandy soils near running streams and stagnant water. It is a low, rather irregular-growing shrub with zigzag branches and almost spinose branchlets. The bark is inclined to a reddish hue, especially when the plants are young. The flowers appear in May, are white and very fragrant, and produced in such profusion as to cover all the branches. They are followed by abundant fruit which hangs long, is nearly an inch in diameter, and is edible. As the shrub grows from five to ten feet in height, its value cannot well be overestimated in the section to which it is indigenous, and it may prove of great service in similar locations. It appears to be perfectly hardy, and has a field for usefulness and also as an ornamental plant throughout the West.

CARAGANA—Siberian Pea Tree.

THE caraganas constitute a small class of ornamental shrubs, not very widely known in cultivation, which are both curious and interesting. There are several species and varieties that are worthy of a place

Caragana—Siberian Pea Tree.

in the garden or border wherever a plant of their proportions is desired. Most of them prove sufficiently hardy to withstand our northern winters, and so adapt themselves to a large extent of country. So far as known none of them is indigenous to America, though most of them appear to be much at home among us. They are all easily grown, and not especially particular as to soils and situations.

C. arborescens.—This is a native of Siberia, and is the species longest and best known in America, and under favorable circumstances grows to a height of fifteen to twenty feet. The foliage is compound, consisting of four to six pairs of leaflets of good color which remain through the season. The flowers are pale yellow, and very numerous, so that when in blossom the little tree somewhat resembles the viburnum. These appear in April or May. There is a pendulous form of much smaller growth, and, when height is desired, it can be grafted on taller stems with good results, as the smaller branches fall gracefully to the ground on every side.

C. altagana is but a small shrub, two to four feet high, with six to eight pairs of leaflets which are glabrous and nearly round. Its blossoms are also yellow. *C. chamlagu* is a native of China, and differs but little from the above except that its flowers appear in June and are first golden and then red. *C. frutescens* is another native of Siberia, and has deep golden-yellow pea-shaped flowers in great abundance, appearing in May and June. It is usually two to four feet in height, and is one of the most valuable of its class. *C. spinosa* is also a Siberian plant,

Ornamental Shrubs.

growing from four to six feet high, with yellow blossoms in great abundance. This is pronounced, on account of its long branches and strong thorns, a most excellent shrub for forming impenetrable hedges. It is especially adapted to sandy soils and to grafting on *C. arborescens* when a taller plant is desired.

ELÆAGNUS.

THE elæagnus is the typical species of the natural order *Elæagnaceæ*, and constitutes an interesting family of deciduous and evergreen shrubs, mostly small trees, which are now coming into quite general use in ornamental planting. Special attention has been called to this group within the past few years through the introduction of several new varieties from the Orient, though one or two of the American species are possessed of valuable qualities, and are not to be overlooked. Nearly all have proved themselves hardy throughout the north temperate zone, while in the southern portion of the United States some of them are nearly or quite evergreen.

E. longipes is a Japanese species, and may undoubtedly be counted one of the best yet known to cultivation. It grows from three to five feet in height, with numerous slender branches which are covered with brown, rusty scales, but are not such as to give it an offensive appearance. Professor Sargent, who saw it in Japan, says that in old age it there attains a height of from twenty to twenty-five feet, and forms a stout trunk a foot in diameter. The leaves are thick, dark green above and silvery-

ELÆAGNUS LONGIPES.

white beneath. The flowers are rather inconspicuous, but the fruit is showy and ornamental. It is borne on long stalks, and is bright red covered with small white dots. It hangs long on the stem, and affords a beautiful contrast to the coloring of the foliage. This elæagnus may well be grown for the fruit alone, as it is juicy and edible, having a sharp, pungent, and agreeable flavor. Though all persons do not enjoy the taste, some esteem it preferable to that of the currant or gooseberry. In France it is used for preserving, and is highly appreciated. The plant is sure to prove a valuable acquisition and to come into more general use. So far as tested the shrub is found perfectly hardy as far north as New England, and even Canada.

E. canadensis is a native of America, and in some sections is popularly known as the Missouri silver-tree. It grows to the height of eight or ten feet, often throwing out an abundance of suckers. The leaves are oblong, sharp-pointed, and silvery-white both above and beneath, contrasting strongly with the yellow flowers which appear in July and August. These are followed by roundish, ovate fruit, ribbed and covered with white scales. The flowers are fragrant, and the fruit, though dry and mealy, is esteemed by many.

E. angustifolia is a native of southeastern Europe and western Asia, and is the veritable wild olive of the classic authors. It is often in modern times called the Jerusalem willow, though it is not a willow at all. The flowers are yellow, appear in midsummer, and the oblong, light-colored fruit is common in the markets of the East.

ELÆAGNUS HORTENSIS—EUROPEAN OLEASTER.

George Nicholson says: " Under cultivation I have seen this thrive in a dry, hungry, sandy soil, and attain tree-like proportions, with a stem as much as a foot in diameter. This deciduous species is capable of being turned to good account by the landscape gardener." *E. hortensis* is a native of mid-Asia, where it grows freely on the mountains, often at an elevation of three thousand feet and upwards. It is there largely cultivated by the natives in orchards, for its fruit. It is scarcely known in this country, but might possibly be introduced to advantage. A report from South Dakota is to the effect that it flourishes in that section, where the thermometer sometimes registers thirty degrees below zero, and where the annual rainfall does not exceed twenty-two inches. The foliage is late in breaking out, so that it escapes late frosts, and the roots go deeply into the earth, thus enabling it to withstand periods when the rainfall is so light that many other sorts fail. It may be used for low hedges, as the lower branches are well preserved, thus constituting an effective windbreak.

E. umbellatus has sometimes by nurserymen and others been confounded with *E. longipes*, and sold as such; but it is not the same. The foliage differs but slightly from that of the type, having the silvery cast, but the fruit is the color of amber and speckled with white, and if possible is still more abundant. It is about the size of a large currant, and fully as valuable. When first ripened it is quite acid to the taste, but a little later it becomes sweet and mellow. Those who know it best speak highly of its value for cooking purposes, and it is likely to come into common

Elæagnus.

use as soon as its merits are fully appreciated. Very few shrubs are more beautiful on the lawn or in the border,

ELÆGANUS UMBELLATUS.

especially in autumn ; and as its fruit ripens in November, when berries of all kinds are scarce, this is certain to serve a good purpose in supplying the deficiency.

E. macrophylla is quite distinct from either of those named, and, in fact, from almost any other shrub known to cultivation, and capable of serving a useful purpose in the garden or shrubbery. It is said that English gardeners are now making free use of it whenever it can be procured, but it has been so recently introduced that the market supplies are limited. It is an evergreen species with large round leaves light gray on the upper surface and nearly pure white beneath. The flowers are greenish-yellow, appearing in clusters of considerable size in early autumn. It is not known that the fruit is of especial value, or that it is perfectly hardy in the Northern States. *E. simoni* is another of the Chinese sorts, less ornamental than several of the others, but has thrown off a variety with colored foliage of great beauty. In this the leaves are margined with dark green, have golden-yellow centres shaded into brown, and maintain these peculiarities almost the entire season. This variety is said to have originated in Belgium, where it is looked upon as a most valuable acquisition. It is doubtful if it has yet been introduced to American gardens, but in the South it would assuredly prove a success.

There are several other evergreen species, classified as *E. glabra*, *E. pungens*, and *E. reflexa*, which so closely resemble each other as scarcely to be entitled to a separate description in a work of this kind. These are all small shrubs, six to ten feet, and have variegated forms of great beauty, and are especially adapted to planting in the Middle and Southern States, as their hardiness may not be sufficient to endure the rigor of climatic conditions in New England and the Northwest.

CAMELLIA.

THIS is a genus of elegant and most interesting plants suitable to southern cultivation only. It belongs to the order *Ternstrœmiaceæ*, and was named in honor of a Jesuit missionary by the name of Camellus, who wrote a history of the plants of Luzon and some others of the Philippine Islands during the last century. Most of the species are tropical or sub-tropical

Camellia.

plants, though several are sufficiently hardy for out-of-door cultivation in or near the Gulf States, where they are coming to be more freely planted from year to year, as their merits are better appreciated. None of them can be successfully grown in the North, but occasional specimens are to be found in the Middle States under favorable conditions and with slight protection. Mr. P. J. Berckmans of Augusta, Georgia, writes that "camellias abound in all the southern cities, where some have reached to a great size and have stood every extreme of heat and cold. Beginning with the old *alba plena*, or double white, whose flowers often open in November, we have a regular succession of floral harvest until April, and have the choice of some two hundred varieties." The same authority says that the best seasons for transplanting are from early October to the beginning of November, and from the end of February to the end of March.

C. japonica is most prominently noted among the species as the common camellia, and is the type from which has sprung a great number of varieties and hybrids now widely distributed. Under the most favorable conditions, it grows in somewhat tree form to a considerable size, and is possessed of great beauty. The leaves are quite large, ovate, sharp-pointed, serrated, and of good substance and color. The blossoms are in numerous shades, and mostly pink with yellow projecting stamens. It is a native of China as well as of several other Asiatic countries. *C. japonica alba* is much the same except in the color of its petals which are of the purest white, contrasting strongly with the bright yellow centres. The original forms are now seldom found in cultivation either in the garden or conservatory, as some of the varieties are much superior.

It would be useless to attempt to describe in detail the almost numberless forms produced through natural processes, and by the hybridizers who have turned their attention in this direction. There are varieties with double and single flowers, and almost every possible shade of color, and the number is increasing every year. It is claimed that there is no plant that will afford such a wealth of bloom extended over so long a time as the camellia, and the claim appears to be pretty well founded. It is essentially a winter-blooming plant, and among all the broad-leaved evergreens is one of the best. It is a pity that only a narrow fringe of our territory is capable of producing the camellias in the open ground.

ITEA.

THE iteas belong to the order *Saxifragaceæ*, and are very little known in cultivation. Though the genus contains only five species, it is distributed freely throughout the United States, China, Japan, the Himalayas, and some of the islands within the tropics. One form only is of interest to us, and that has recently been rescued from the long list of neglected plants so many of which are now coming into notice. The botanists call it *Itea virginica*, and so far as known this is the one species looked upon with favor in the parks and gardens. It grows to a height of five to eight feet, and has alternate leaves, oblong, pointed, and minutely serrate, changing from green to scarlet-crimson in midsummer, and retaining the new shade until autumn. The flowers are white, and in terminal racemes, small, but sufficiently numerous to make the bush attractive, and to justify a greater use

ITEA VIRGINICA

of the plant than it now has. They appear in the middle of June. Though supposed to be a distinctively southern plant, the itea is found to grow wild in New Jersey, and may be safely used much farther north.

VITEX—Chaste Tree.

OF the vitex there are only one or two species sufficiently hardy to withstand the rigors of our northern climate. The genus is of the order *Verbenaceæ*, and contains not far from sixty species, mostly classed as greenhouse or stove plants. *V. agnus-castus*, though supposed to be too tender for garden cultivation north of Washington or Philadelphia, is, nevertheless, occasionally seen in New York and even in southern New England, but in such locations it demands and must receive thorough protection. Thomas Meehan writes that in Philadelphia it gets partly winter-killed, but that this does not hurt it in the slightest. On the contrary, the shoots seem to start more vigorously from the base, and to give finer flowers than they otherwise would. Where it does not winter-kill it would be well to cut the plant to near the ground, as is done with the hydrangea. The vitex is popularly known as the chaste tree, though also bearing such names as hemp tree and pepper tree, the latter perhaps in recognition of the fact that its foliage gives off a peculiar, aromatic fragrance by no means objectionable. It is a small, neat-appearing shrub, from three to six feet in height, and with long, narrow-pointed leaves, and panicles of bright lilac flowers shooting up above the foliage. They are especially welcome, as they appear in August and September when flowering shrubs are not common. There is a variety of vitex having white blossoms, and another with deep blue ; and a species recently introduced from China is said to be more hardy, but this is not yet sufficiently tested to disclose its full merits.

CORNUS—Cornel—Dogwood.

THE cornus family, of the order *Cornaceæ*, includes trees, shrubs, and a few perennial herbs widely distributed through Europe, Asia, and America. The Latin name comes from *cornu*, the horn, and was applied because of the hardness and strength of the wood, and its real or supposed durability under exposures. The

Cornus—Cornel—Dogwood. 277

garden varieties are among the most valuable shrubs and small trees in use. It is in their favor that they will do better in the shade and when exposed to the drip of overhanging trees than do most other shrubs. For this reason, if for no other, they fill an important place, and, being compact and bushy, they quickly supply vacancies and cover the naked stems of trees or other objects which are desired to be hidden from the eye. All the woody species can be used in this way to especial advantage. In some parts of Europe, especially in Italy, they have been planted for hedges, and with satisfactory results. They are nearly all remarkably hardy, adapting themselves readily to great diversities of soil and climate. All are deciduous, and mostly with leaves opposite, entire, and of good substance. Some of them produce flowers of great beauty, and in abundance. In most cases the bark is bitter and astringent, as are also the berries, which ripen in autumn. The wood is close-grained, and much prized for purposes requiring strength and endurance.

C. florida, flowering dogwood, is one of the most desirable of our native shrubs. It usually grows from ten to twelve feet high, but occasionally, under favorable circumstances, shoots up to twenty or thirty. In either case it assumes fair proportions, and in its season of blossoming, which is early spring, commands universal admiration. The leaves are four or five inches long, and two to three wide, ovate, sharp-pointed, and somewhat pubescent or hairy, especially along the mid-rib and more prominent veins. The flowers appear at the end of the branches, twelve or more in a head, and are supported by short, club-like stalks. They are small and attract but little attention, and what is taken as the blossom is the

CORNUS FLORIDA.

whorl of large leaves by which the real flowers are surrounded. These are four in number, pure white, and spreading so as to become very conspicuous, and whether seen in the edges of a forest or in the garden border are very beautiful. The fruit which follows is closely bunched, bright scarlet, and also showy. It is so bitter that even the birds will not touch it until its character has been somewhat changed by frost, when it becomes acceptable to the robins, and probably to the voracious little sparrows that are always with us. The bark of the stem and branches is also very bitter, and is sometimes used successfully as a substitute for Peruvian bark as a tonic, an astringent, or an antiseptic. In autumn the foliage changes to purple and crimson, and with the bunches of crimson berries makes the tree almost as attractive at that season as it was in springtime. Like most of the cornels, it is of slow growth and entirely hardy.

Some of the earlier botanists who visited this country and came in contact with this shrub or tree, were enthusiastic in its praise, as well they might be. William Bartram, in his *Travels in Georgia and Florida*, gives the following account of its appearance as he found it near the banks of the Alabama River: " We now entered a remarkable grove of dogwood trees, which continued nine or ten miles unaltered, except here and there by a towering *Magnolia grandiflora*. The land on which they grow is an exact level; the surface a shallow, loose, black mould, on a stratum of stiff, yellowish clay. These trees were about twelve feet high, spreading horizontally; and their limbs, meeting and interlocking with each other, formed one vast, shady, cool grove, so dense and humid as to exclude the sunbeams, and prevent the intrusion of almost every vegetable; affording us a most desirable shelter

280   Ornamental Shrubs.

from the fervid sunbeams at noonday. This admirable grove, by way of eminence, has acquired the name of The Dogwoods. During a progress of nearly seventy miles

FLOWERING DOGWOOD.

through this high forest there was constantly presented to view, on one hand or the other, spacious groves of this flowering tree, which must in the spring season, when covered with blossoms, exhibit a most pleasing scene." And Professor E. L. Greene in our own time says: "One of the delightful, unfading pictures in our memory of eastern woods in their June glory is that of the shrub or small tree known as flowering dogwood. A full-grown specimen with its widespread and stratified branches, each ultimate twig bearing a large, white, cruciform involucre, which commonly passes for a corolla, is an object of striking beauty in the finest glades where it occurs."

The species has several varieties of value in cultivation. Of these *C. f. pendula*, or weeping dogwood, is one of the most striking, having foliage and flowers like its parent, but borne on pendulous branches on every side of the upright stem, and extending to the ground. The branches are firm and rather stiff, though not always so represented in the pictures shown in the nurserymen's catalogues. The fruit consists of red berries, and in autumn the foliage changes almost to crimson. *C. f. rubra*, or the red-flowering variety, is of quite recent introduction. It is much the same as the type, except in the color of the blossom, which may be described as deep rose, dark red, or sometimes pink instead of white. The original is said to have been discovered on one of the Virginia mountains by a clergyman, through whom it was domesticated and introduced to civilized society. Though interesting and worthy of extensive planting, it is not more beautiful than the type. It makes a fine single plant, but perhaps the best use to which it can be put is to place it in the shrubbery beside the older forms, when the contrast cannot fail to be especially agreeable.

C. circinata, round-leaved dogwood, is an American species, growing in moist situations from Canada to Florida. It is from five to ten feet high, with numerous slender branches which form a well-shaped and spreading head. The bark is usually marked with warty dots which fail to add to its beauty, though not especially objectionable. The leaves are oval, abruptly pointed, prominently veined, somewhat rough, and covered with a whitish bloom or down beneath. The flowers are white, in terminal, flat cymes, on bowing footstalks, and appear in May. The fruit is small, pale blue, ripening in early autumn and remaining until after frost. While not one of the best of the cornels, it proves an acceptable plant for the shrubbery or border. Unlike most of the species, it prefers a rocky or sandy soil, and proves of good service for such situations.

C. alternifolia is found chiefly in the Middle and Western States. It grows sometimes as a tree twenty to thirty feet high, with branches somewhat in whorls and quite numerous; but more frequently appears as a shrub eight to ten feet high, and of good form and character. Unlike most of the dogwoods, both its branches and its leaves are alternate. In the case of the former, the bark is of bright, shiny green with splashes of gray varying considerably in size and form. The leaves are borne on long footstalks, are wedge-shaped at the base, and lanceolate toward the apex, terminating in a rather sharp point. They are dark green above and almost glossy, but lighter beneath and slightly pubescent. The flowers, which appear about the first of June, are in somewhat irregular clusters, white, though sometimes tinted with yellow. The fruit is black tinted with blue, and in its abundance adds materially to the value of the plant. It is less bitter and astringent than that of some of the other species, and quickly taken by the birds.

C. stolonifera, the red osier or red-stemmed dogwood, is a well-known shrub found growing freely in wet, marshy lands throughout Canada and most of the Northern States. Its main stem is prostrate, and wholly or partially under ground, whence it throws up an abundance of small, straight shoots six or eight or even ten feet in height. These are clothed with a smooth bark, dull green or reddish in summer, but becoming a glowing scarlet in winter. This is, in fact, its chief attraction, as the show of color is in marked contrast with surrounding objects. In some instances a mass of these shrubs at that season looks almost like a sheet of fire, when seen at a little distance. In summer the stems throw out large, roundish leaves, somewhat rough on both sides, and terminating in a short, sharp point. This species appears to be also indigenous to Europe and Asia, growing as far north as Siberia, and, as might be expected, is entirely hardy. It is there known as *Cornus alba*, with reference to its fruit, which is a small berry, bluish white and very bitter.

C. sanguinea, another red-stemmed cornus, is a species of foreign nativity, though now well known and common in American gardens. The height is from six to eight feet, with numerous shoots proceeding from a more or less prostrate root, which is in reality the stem partially covered by the soil and an accumulation of leaves. The greenish-white flowers appear late in spring, but are not especially attractive or interesting. The foliage is good, making it a valuable shrub for summer, though its chief attractiveness is in winter when its red branches render it conspicuous. The fruit is a small black berry, which, when compressed, yields a valuable oil that in some countries is used in the manufacture of fancy soaps and other articles requiring oleaginous substances. The species is very abundant in western Asia and some portions of Europe.

C. spathi is a variety in many respects decidedly superior to the type, and one of the very best variegated plants. The leaves in spring are dark, almost bronze, and very attractive; but the greatest charm is put on in midsummer, when they are broadly and irregularly margined with yellow and white, which is retained the remainder of the season, the peculiar shading being more constant than in many of our most popular colored foliage-plants. Though brighter in the open sunshine, the variegation is well maintained in partial shade. It is scarcely possible to find a more strikingly beautiful and charming shrub for planting singly, in masses, or in promiscuous groups and borders.

C. sericea, or silky cornus, is a somewhat spreading shrub growing freely on the banks of streams and in moist places, seldom reaching above five to ten feet in height. It produces white flowers in corymbs during the months of June and July, and these are followed by pale blue, globose berries. The younger branches are somewhat purple, sprinkled with white and covered with a silky down, whence comes the name. The leaves are opposite, two to three inches long, sharply ovate, rounded at the base and pointed at the apex. This cornus is a native of the United States, and ranges from Canada to the Gulf of Mexico, having a preference for moist lands, though growing well in any reasonably good soil. *C. paniculata*, or panicled dogwood, produces its white blossoms in loose cymes or panicles in July and August, much later than most of the other forms, and for that reason is especially desirable. It grows from five to seven feet high under ordinary conditions, in bushy form and with whitish leaves, and has berries in late autumn.

C. mas, or cornelian cherry, is a native of central Europe, but, though introduced to America many years ago, is not as often seen in our gardens as it should be. It is a small tree or large shrub, reaching sometimes the

Cornus—Cornel—Dogwood.

height of about fifteen feet, with slender branches and a well-rounded head. Its chief beauty consists in the small, bright yellow flowers which appear in early spring in advance of the foliage. These blossoms are in compact clusters which extend the whole length of the branches, giving the tree a very striking appearance. As it is one of our very first bloomers, it occupies an important place in ornamental planting. There are two varieties of this plant which are especially beautiful. One, *C. m. variegata*, has its foliage strikingly marked with pure white, and the other, *C. m. elegantissima*, is white with light and yellow shadings. Both are desirable garden plants. Thomas Meehan in writing of this plant takes occasion to say: "In this dogwood we can see how Nature makes species! We are all familiar with the white dogwood of the woods,—*Cornus florida* in the east, and *Cornus nuttallii* on the Pacific slope. We know well the four broad, white bracts which lend the dogwood flower its chief charm. We see in the cornelian cherry the same four bracts, only that more correctly they are the scales that protected the flowers in

CORNUS ELEGANTISSIMA.

the winter. In the other dogwood cited, the four bud-scales simply take a second growth, carrying the winter portion on the apex. The notch on the end of the broad, white bract is the bud-scale of the past winter. What the power is that says to the bud-scales of *Cornus mas*, 'Rest when you let the flowers out,' and to the bud-scales of *Cornus florida*, 'Take another growth and become another species,' nobody knows yet,—but it certainly is not by any law of natural selection, the struggle for life, the survival of the fittest, or the accident of environment. There are many reasons why this lovely shrub should have a place in any garden that can find room for it."

C. stricta, or upright dogwood, grows to a height of eight to fifteen feet, with numerous straight stems or branches, making a beautiful bush. The blossoms are white, in open cymes, and showy; the foliage is bright green. One of its varieties, little known, has its leaves beautifully variegated with yellow and white, but the colors are not as permanent as might be desired. *C. baileyi* has scarcely been introduced to cultivation, but it appears to possess qualities that should make it useful. It grows freely in certain sections of the Northwest along the borders of the Great Lakes, and, according to Professor Bailey, on sand-dunes, and often in the loosest, shifting, white sands. Any plant which will do this is capable of great service in many localities, and especially a shrub of upright form, good foliage, and beautiful flowers which appear continuously from June to September. The fruit of this cornus is in clusters of pearly white, and is quite showy.

C. kousa has come to be known as a cornus, though it was formerly classed as *Benthamia japonica*, that genus having now been merged into this. The *kousa* is an interesting form, having flowers which are yellow, very small, and borne in clusters, the showy part of the inflorescence being furnished by four large white bracts which surround the real blossoms exactly as in *Cornus florida*. The bracts of this Japanese cornus are, however, more pointed and, if anything, of a purer white. Mr. Falconer, who grew it at Glen Cove, pronounced it one of the finest shrubs one could have in a garden, and far more hardy than some of the other Japanese sorts.

GORDONIA.

THIS is a distinctively American tree or shrub, having been discovered on the banks of the Altamaha River in Georgia, by a botanist of the last century. It is said to have been named in honor of Dr. James Gordon of Aberdeen, with whom the discoverer had previously been associated. It belongs to the order *Ternstrœmiaceæ*. The genus includes few species sufficiently hardy for out-of-door cultivation, and even in the middle-southern States winter protection is sometimes needed. With suitable precautions specimens have been grown as far north as New York and even Boston, but the gordonia cannot be advised for planting, except by experts, much above Washington.

Although indigenous to the United States, and here first found, it is not now known to be growing wild in any part of the country. There are only two forms worthy of especial mention as suitable for the park or garden. The first, *G. lasianthus*, popularly known as loblolly bay, is a shrub often rising to a height of eight or ten feet, with coriaceous foliage, the leaves being oblong-lanceolate, narrowed at the base, smooth and glossy. The flowers are about four inches in diameter, pure white, and deliciously fragrant. They are composed of five broad, incurved petals enclosing a large number of yellow stamens and anthers. The flowers appear early in September, and, though they are at no time abundant, continue in succession until killed off by frost. This late blooming gives the shrub its principal attraction. The whole bush has a peculiar fragrance said to resemble that of the Chinese tea plant, so much so that the leaves have sometimes been used as a substitute for tea. *G. pubescens* is much the same in its general characteristics. It grows to little more than half the height of the preceding, and has leaves slightly downy, especially on the under side. The flowers are about three inches in diameter, pure white with yellow filaments, and fragrant. They appear in August, and continue until late autumn.

CERCIS—Judas-Tree—Red-Bud.

WHAT is popularly known as the Judas-tree or red-bud, *cercis*, constitutes a genus of *Leguminosæ* containing, so far as known, not more than five or six species, and only a small number of varieties that are found in cultivation. The common name, Judas-tree, is applied because of the ancient legend that the arch-traitor went out and hanged himself on a tree of

Cercis—Judas-Tree—Red-Bud.

this class, reference being doubtless had to the appearance of the tree in blossom, when its trunk and branches are covered with small buds or flowers much resembling drops of blood. There is another legend to the effect that the ignominy fell to the lot of the elder, which has ever since had a repulsive odor. The family is indigenous to the south of Europe, to Eastern Asia, and to North America. It is distinguished among the order to which it belongs by its glabrous, kidney-shaped leaves, and the peculiar flowers, to which reference has been made. These buds and blossoms are succeeded by thin, flat, brown pods sometimes nearly six inches long, remaining on the tree nearly or quite all the year. In garden cultivation, especially in colder climates, these are seldom followed by fruit, resort being made to other methods of propagation for the perpetuation and increase of the family.

C. canadensis — American Red-Bud.—This is the species best known in our parks and gardens. It is a fine tree in the early season, showy from the appearance of its buds, which break out in great numbers along nearly the whole length of its branches, and are of a brilliant red or rose color. Nothing among trees is more singular and attractive, and when a well-formed top is thus ablaze one can scarcely pass without pausing to admire. The tree is not large, though often twenty-five to thirty feet high. The leaves, which begin to appear in May while the flowers are expanding, are folded in a peculiar manner on the bud, and when fully grown are somewhat heart-shaped at the base, acuminate, and of a deep, rich color.

C. siliquastrum is a native of southern Europe and

various Asiatic countries, and has been longest kwnon in cultivation, having had a place in famous gardens for many generations. It produces slightly larger flowers and of a somewhat darker shade than does the American species, but they are not more beautiful, nor are they brought out in equal profusion. It is not found to endure our northern winters so well as the native sort, but may be safely planted throughout the Middle and Southern States. The foliage is quite obtuse, and nearly circular, but in its general characteristics equally good.

C. chinensis is the eastern species, and was received in this country through the medium of Japan, and has come to be popularly known as the Japan Judas-tree, though it was probably not indigenous to the Island Empire. It has larger leaves and flowers, the latter appearing, if possible, in even greater profusion than on its western congeners. When at its best, it appears as a perfect sheet of flame, and by the branches spreading the effect is heightened to a marvellous degree. In growth it is smaller than either of the other sorts, and more shrub-like, seldom attaining a height above twelve feet. It is probably the best of its class. No garden, park, or lawn is complete in its spring exhibit without a red-bud or Judas-tree in one or more of these forms.

C. texensis has its home in semi-tropical climates, and is distinctively a tree for planting in the South and not in the North. It is met with most frequently in the State from which it takes its name, and will doubtless do good service along the lines of latitude suggested. Its resemblance to *C. canadensis* is so close that the botanists have not been agreed as to whether it is a distinct species, or a mere variety of the better-known form. But if it answers the purpose for the South which the common Judas-tree does for the North it will prove an acquisition to the parks and gardens of that section. It is described as a slender tree from twenty to thirty feet high at maturity. As with the other members of the family, the foliage and flowers are put forth nearly together, and in early spring. The flowers are about half an inch long, and are on slender pedicels, though the clusters are nearly or quite sessile.

CORYLOPSIS.

ONLY a few species of corylopsis appear to be known in cultivation in America, though there are several that Nicholson in the *Dictionary of Gardening* pronounces very ornamental and interesting, hardy, deciduous shrubs. None of them is native of the western continent or of Europe. So far as known they all come from China or Japan, with the exception of a single specimen from the Himalayas. The genus at best is a small one, but the wonder is that it has not been more taken into cultivation, both here and in England, where it is just beginning to make its way.

C. spicata is the best-known form. It is a small bush three to four feet high, and was introduced from Japan, where it is grown in the best gardens, and highly esteemed. It has long-stalked, feather-veined foliage, finely serrated, and glaucous beneath, smooth and pale green above. The flowers have five petals and five stamens, and spring from the axils of yellowish-green bracts, and are disposed in racemes two to five inches long. They are pale yellow or lemon in color, and those familiar with them detect the odor of cowslips as nearly as it can be defined in words. These flowers appear before the leaves, and sometimes open in midwinter when the weather is mild for even a brief period. The plant needs to be cut back severely in transplanting, and in the extreme North winter protection will be of service. *C. pauciflora* is much the same in its general characteristics, but has fewer and smaller flowers, and does not attain to so large a size. *C. multiflora* comes from the tea districts of China, and is also an interesting plant. It has more rigid leaves less distinctly veined than the *spicata*, and grayish beneath, and longer and more closely packed racemes of yellow blossoms. The odor is something like that of the tea plant. This species has been so recently introduced that it is scarcely known to the nurserymen. *C. himalaya* is another promising sort, having lighter-colored blossoms and still longer racemes, but it has not been sufficiently tested as to its climatic range to be advised for general planting.

HAMAMELIS—Witch Hazel.

THOUGH in the highest sense the witch hazel, as known under ordinary conditions, is scarcely to be included in the list of ornamental shrubs, it is nevertheless an interesting plant and capable of good service. There are only three known species, with perhaps two or three varieties, and these are nowhere largely in use as garden plants. One of the species is of American origin, one probably a native of China, and the third of Japan. In this country *H. virginica* is best known, as it grows freely over a large portion of our extended domain. It rises at its best from twenty to thirty feet, but is usually of much smaller dimensions. In most cases it assumes a bushy form with several stems springing from a common root, each branching freely so as to form a somewhat open and broad head. Occasionally it takes the form of a small tree, branching near the ground so as to still give it a shrubby appearance. Its chief peculiarity is in its flowers and fruit. The former are gathered in axillary clusters of three or four, are bright yellow, and, though small, especially interesting from both a botanical and horticultural standpoint. Nicholson, in his *Dictionary of Gardening*, says: "During the autumn and winter they begin to expand before the leaves of the previous summer drop off, and continue on the bush through the winter; after the petals drop off in the spring the persistent calyces remain until the leaves re-appear in April or May." This is the English description, and it answers to what is known of the plant at home. It is no uncommon thing to see these blossoms at any time between October and March, as the bush is found skirting the forest or growing along the banks of brooklets from New England to Texas. They are followed by two-celled, woody pods, each containing a small nut which is edible and quite agreeable to the taste. The pods mature late in the following season, and often not until flowering time. The leaves contain a large amount of tannin, and the product is used as an astringent and for other medical purposes, though chemists fail to find in it elements for the cure of the many ills for which the decoction is recommended.

CALOPHACA.

Calophaca wolgarica is a small shrub but little known. It belongs to the lentil family, and comes from Siberia, having been introduced as long ago as 1780 to English and continental gardens, where it has since only barely held its own. It grows to a height of about three feet, and has the advantage of being reliably hardy and of good form, with pleasantly-tinted, pinnate foliage, and abundant golden flowers in hanging racemes, affording an agreeable contrast to its leafage. To bring this out to the best advantage it is advised to graft the more humble plant on the laburnum, and at such a height as may be desired, when, says Nicholson, " it forms an object at once singular, picturesque, and beautiful, whether covered with blossoms or with its fine, reddish pods." As a low plant it serves an excellent purpose in the edges of borders as well as in groups or masses, and when given sufficient height, as suggested, it surpasses many popular favorites. It thrives best in rather dry soil and partial shade. *Cytisus wolgarica* and *Cytisus pinnatus* are synonyms under which it is sometimes catalogued.

PHILADELPHUS—Syringa—Mock Orange.

THIS is a small genus of some twelve or fifteen species of the order *Saxifragaceæ*, indigenous to southern Europe, central Asia, Japan, and North America, all hardy shrubs, and possessed of many quali-

Philadelphus—Syringa—Mock Orange.

ties that commend them for garden cultivation throughout the temperate zones, where alone they are supposed to thrive. Without exception they are of easy cultivation, and few shrubs make so good returns, in both foliage and blossom, for the outlay expended upon them. As the flowers are produced on wood of the last year's growth, it is well to cut the shrub back sharply immediately after the flowering season. If this is not done it is liable to become a straggling bush, bare near the main stem, and somewhat coarse. By cutting, the number of flowering branches is multiplied, and, though the bush becomes large, every part will prove floriferous. The popular name, syringa, should be abandoned, as that belongs to the lilac. Philadelphus is as easily remembered, and it is better to call things by the right names.

The best known of the species is *P. coronarius*, and there is reason for speaking of it as the mock orange because of the resemblance of the flower, and of the fragrance of the entire shrub, to that of the real orange tree of the South. Under favoring conditions it grows to a height of twelve feet. It is compact if properly handled in cultivation, but if left to itself it is sometimes far from symmetrical, though never ugly. The leaves are ovate, sharply pointed, and serrate. The flowers are creamy-white or light straw-color, and are possessed of a pungent fragrance. They appear in May, and in great abundance. There are several varieties, one of which, *flore pleno*, has double flowers of the same color and with similar fragrance. Another, *argentia marginata*, has the foliage bordered with white, and is quite distinct and beautiful. It is a recent introduction, and is not yet widely distributed. For limited grounds it may be preferred, as it is of somewhat smaller habit and more compact. If a real dwarf is wanted, the variety *nanus* may be chosen, though it does not blossom quite as freely. *P. c. zeypheri* is, on the contrary, a large, spreading bush, and has beautiful flowers

PHILADELPHUS SPECIOSISSIMUS.

without fragrance, and which appear much later. What is popularly known as the golden syringa, *P. folius aureus*, has distinct yellow leaves, and is in all respects one of our best foliage plants. It holds its color throughout the entire summer, and, whether grown as a single plant or in masses, is a superior small shrub.

P. gordonianus is a native of this country, and, coming

from the Northwest, is entirely hardy. It has ovate, pointed foliage slightly serulate, and with favorable circumstances makes a bush ten to twelve feet high and almost as broad. The flowers are produced in great abundance, are almost scentless, and in terminal racemes of from five to nine blossoms. This is much planted, and is one of the best sorts, as it is in its prime in July after most of the others have gone by. *P. grandiflorus* is another native of the United States, having its home in the South. It grows to about the same proportions, and has nearly round foliage, pubescent in the early part of the season, and irregularly toothed. The flowers are much larger than those of most of the other species, and are possessed of a delightful fragrance, not as pungent as that of the *coronarius* which to some people is offensively strong. One of the varieties, *P. g. speciosissimus*, is of dwarf habit, and especially attractive as a garden plant, particularly where space is a consideration. It grows about three feet in height, and produces in great profusion very large, pure white, fragrant blossoms.

P. microphyllus differs from most of the other species in having small foliage, the individual leaves being from one half to three quarters of an inch long. They are ovate, lanceolate, entire, and numerous. This grows about three feet high. The flowers are large, terminal, solitary, and in threes. As the branches are erect and slender, the plant has a very graceful aspect wherever seen. *P. nivalis* has glaucous leaves, white on the under side and green above. It is a small plant with the customary white flowers of the genus, as is *P. hirsuta*, or the hairy species,

Ornamental Shrubs.

the leaves of which are covered with hairs on both surfaces.

There has been recently introduced a class of hybrids some of which are known to be of a superior value. They are the products of the skill and ingenuity of M. Lemoine, to whom the horticultural world has become so greatly indebted as a hybridist. Most of them are not yet in cultivation in America, but doubtless will be at an early day. Among others they include the following : *P. lemoinei avalanche*, the best known, is described as having long, slender branches, and very large, white, fragrant blossoms, bending the stems under their weight. It is a bush from six to eight feet in height, and has often been figured in the magazines and catalogues, and is more or less familiar. *P. lemoinei candelabre* is another free-flowering variety, and has white flowers of unusual size, and prettily dentated and undulated. This is quite dwarf in habit, and makes a charming, compact little mass of blossoms. *P. lemoinei erectus* is an upright bush, and has small, very sweet flowers, while two others, called *sheaf of snow* and *Mont Blanc* have large, fragrant flowers like the others of this class, completely covering the shrub at the

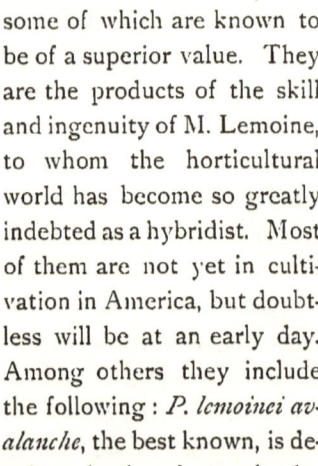

PHILADELPHUS CORONARIUS.

time of blossoming. The flowers of *silver ball* are also sweet-scented, and large, and vary through being double.

ARALIA—Angelica Tree.

THERE are said to be twenty-five or thirty species of the genus aralia, and widely distributed; but they are mostly tender plants, and, except in warm climates, suitable only to cultivation under glass and with artificial heat. Several prove to be half-hardy, and can be grown in the open in such localities as California and the States bordering on the Gulf of Mexico, and northward to the Carolinas. All are interesting and even beautiful plants, and worthy of more attention than has been usually accorded them in American gardens, though but two or three are able to endure the winters of New England and the middle Northwest.

A. spinosa—Angelica Tree—Hercules' Club.—This is one of the largest growing, and the best of all for garden use. It rises from twelve to sixteen feet in height, and has the habit of sending up branches from the roots, so that when once established it often becomes a group, the parent stem in the centre, with smaller and lower ones on every side. When desired, these secondary growths can be easily removed so that the tree form may be retained. This species is indigenous to North America, and when first seen by Europeans was regarded as a great curiosity. One of its peculiarities is that the woody stem is covered from end to end with sharp prickles, so that one can scarcely touch it with the bare hand; and another, that in autumn—especially when young—the large stems, having

Ornamental Shrubs.

served as branches, fall off as well as the leaves. In this state it appears more dead than alive, and often very much unlike its floral or arboreal surroundings. But when spring comes it is quickly reclothed; the ephemeral branches grow rapidly, and send out compound leaves two or three feet long, and often half as broad. These form a cluster at the top of what was so recently a bare stem, and are very tropical in appearance, being twice or thrice pinnate,

ROSE ACACIA. (See page 46.)

and borne on petioles, fifteen to twenty inches long, which clasp the main stem with a thickened and enlarged base. The flowers come out in midsummer in umbels and compound panicles. They are white tinted with green, and in such masses as to create surprise to one not familiar with the plant. This round head, with large foliage and immense cluster of flowers, has much the appearance of a tropical palm in full blossom, if such were a possible thing.

Aralia—Angelica Tree.

The tree should be planted in sheltered position, as the top is too heavy and too large to withstand high winds to the best advantage. It is well adapted to shady situations, and even prefers them to bright, sunny exposures. With advancing age there is a tendency to more permanent branches and a still larger head. It is a rapid grower, and comparatively indifferent to soils and situation, though when growing wild is found most often in moist locations.

A. sieboldii—Japanese Aralia.—Though belonging to the same class, this is quite another plant from that already described. It grows in the form of a bush three to six feet high and nearly as many through, and is covered with deep, glossy foliage, the individual leaves being digitate, twelve inches across, and on stout petioles a foot long. They are finely cut, and in themselves sufficiently showy to make the plant worthy of a place in our best gardens where the conditions will allow. It is almost an evergreen in the South, where alone in this country it can be successfully planted. The flowers are comparatively small, white, and exceedingly numerous, covering in their

ARALIA FATSIA.

season the umbrella-like head in the greatest profusion. The London *Garden* speaks of it as "one of the most ornamental plants that can be used in the garden, a large mass of it on the turf presenting a fine aspect." It can doubtless be grown to perfection over a considerable section of the United States south of Washington, with the same care and attention bestowed on many other and less worthy kinds; but in the North the winters are too severe. The Japanese name for this species is *fatsia*, and the botanists are now cataloguing it as the *Aralia fatsia*. There are two varieties, one having its foliage marked with white, and the other with yellow, neither proving itself superior in value to the type.

HIBISCUS—Mallow—Althea.

HIBISCUS is the old Greek name applied to what we know as the marshmallow, which constitutes a genus of *Malvaceæ*, having its home mostly in tropical or semi-tropical climates. It includes not far from one hundred species, and almost numberless varieties, some of which are annuals, others herbaceous perennials, and still others large and vigorous-growing shrubs. In nearly all, the flowers are large and beautiful, so that the genus is one of the most desirable in cultivation. Except in a few instances, they are of value in the temperate zones only as stove and greenhouse plants, though there are several, accounted tender, which do good service in open ground as we approach the southern boundaries of the great republic. In a general way they nearly all resemble in blossom the common hollyhock of the gardens,

Hibiscus—Mallow—Althea.

to which they are in reality closely allied. The best-known native species is *H. moscheutos*, found chiefly in swamps and marshes along the Atlantic coast, and to some extent on the borders of the Great Lakes as far north as the Canadian line. This is known almost everywhere as the swamp rose, and with a considerable degree of propriety. When transferred to ordinary soil it continues to thrive nearly as well as before. This class of mallows appears in many colors. The several varieties make charming plants in the border, especially as they flower in mid-summer.

H. syriacus is the bush or small tree popularly known as the althea, and sometimes as the rose of Sharon.

HIBISCUS.

It is believed to have originated in Syria, though it may not have been the flower mentioned as with affection in the Jewish Scriptures. It is usually a shrub ten to twelve feet in height, but sometimes much larger. Its habit is rather stiff and straggling, and it needs attention and a somewhat free use of the pruning-knife to keep it in good shape. In no case is the foliage particularly attractive, though in appearance it is always fairly good. The flowers, double and single, take on a considerable range of form and color, red, white, and purple, with numerous combinations and intermixtures. They come forward at the season of the year when the gardens and fields are most destitute of color, blooming in August and September, and continuing for some weeks. Because of their submission to the shears, the altheas are capable of making beautiful flowering hedges, though they do not appear to be much used for that purpose—perhaps, in part, for the reason that they afford but slight protection as a barrier against encroachments on the part of small animals. But as single specimens or in borders, they show to better advantage than any other shrub at the particular period when they are at their best. The many forms known to cultivation are supposed to be varieties springing from a common type, and as a rule are greatly superior to the species from which they have sprung.

Of the numerous varieties only a few need be named, such as are believed to cover the field of usefulness on the part of the most exacting planters. One of the best-known single forms is *H. totus albus*, with pure white, single blossoms of large size, and without the prominent crimson eye for which the group is distinguished. *H. alba plena* is equally good in its general characteristics, producing large, white, double blossoms. *H. boule de feu* is in two forms, one with single red or pink flowers, and the other with double blossoms of the same hue. *H. alba*

Hibiscus—Mallow—Althea.   305

variegata is marked with stripes of white and red. *H. cærulea* has large, double pink and white flowers shaded with purple, while *H. leopoldii*, which is comparatively new, presents the peculiarity of attractive, cut-leaved foliage and very large as well as double flesh-colored flowers shaded with rose.

H. anemonæflora is one of the later introductions, and proves of especial merit. The flowers are double, the stamen-petals making a tufted centre and much shorter than the true petals. These latter are broad and conspicuous, deep scarlet at the base, distinctly extending into veins to almost the edges of the petals. The general color is dark pink. The plant is noted as a later bloomer than most others. *H. camelliæflora* is as desirable as the preceding through its distinction in color. The large, white flowers are handsome and showy, the broad, true petals extending much beyond the inner stamen-petals. The rich crimson eye, occurring in almost all the varieties, is very marked in this case. *H. pæoniæflora* produces a blossom resembling a peony, though not as large. In color it is a beautiful pink, and very showy. Meehan in his *Monthly* gives the following description of an entirely new form: " *H. enchantress* is practically unknown to the public, not

DOUBLE-FLOWERING ALTHEA.

having yet been disseminated. Its flowers are single, white with the usual crimson eye, and with a dash of rose tipping each petal. The appearance of this color regularly located on the petals, is unique and pleasing." Buist's variegated althea is distinguished for its beautiful foliage. The leaves are marked with stripes of yellowish-white in such proportions and distinctness as to arrest the attention of every passer. It is a low, well-rounded bush, and worthy of a place in every garden or border. It can scarcely be called a flowering shrub, for the blossoms fail to materialize. Though the petals appear to be fully formed, they gather in the shape of balls or buttons, simply showing their tips, which are red. They remain into late autumn, but are more curious than beautiful.

PAVIA—Dwarf Horse-Chestnut.

THE common horse-chestnut, so freely planted in this country, is of foreign origin, having come from China, with Europe as an intermediary, and it must be confessed that America has given it a right royal welcome. Few trees are more freely grown on strictly ornamental grounds, as it has little or no economic value either in its timber or its fruit. It is a large and magnificent tree, and in spring is one of the most beautiful both in leaf and flower.

It may not be appreciated by all that we have a branch of the horse-chestnut tribe which may be properly classed among shrubs, and which is equally, even if not more, desirable than the larger sort. The *Æsculus parviflora*, known also as *macrostachya* in the catalogues, is a

Pavia—Dwarf Horse-Chestnut.

native of the mountainous sections of the Carolinas, and extends thence both southward and westward. As it proves reasonably hardy throughout the Northern States also, it comes nearer than most others to meeting the demands of all sections. Still it cannot be advised for

ÆSCULUS PARVIFLORA.

the extreme Northwest or the most exposed portions of New England. In the Ohio valley, where it abounds, it is popularly known as one of the buckeyes, and, though never a tree, it gains a height of fifteen to twenty feet, with a broad, well-rounded head, the foliage resembling that of the common horse-chestnut, but sufficiently distinct to be characteristic. The leaves are smaller, and composed of from five to seven oval-obovate leaflets, somewhat rough and hairy on the under surface but smooth above. The flowers are in long, erect spikes resembling in form those of the hyacinth, are mostly white, and in July cover the entire bush, which at that time is one of the most beautiful to be seen on the lawn or in the garden. But it is not to be planted indiscriminately, as it throws up suckers from the roots, forming in a few years larger groups than in many situations might be desired. The proper place for it, therefore, is where there is plenty of room to be devoted to its occupancy, where it may be left to itself in the certainty that it will become a thing of beauty scarcely surpassed in its season by tree or shrub. It has the advantage also of being quite indifferent to soils and ordinary climatic influences, for it will grow in the stiffest clay as freely and as surely as in ordinary garden loam.

P. rubra, known in some sections as the red buckeye, and in others as the red or scarlet horse-chestnut, has elliptic-oblong leaves, tapering to a point at both ends, the leaflets numbering from five to seven, mostly in fives, smooth and slightly hairy in the axils of the nerves beneath. The flowers are produced in long and rather open panicles at the ends of the stems and branches, and in May or early June, and being red never fail to attract attention. The botanists do not agree whether it should be classed as a pavia or a horse-chestnut, and so it is placed in either section as the judgment of the horticulturist may dictate. For this reason it has several synonyms, among which are *Æsculus pavia*, *Æsculus rubicunda*, and *Æsculus carnea*. There are several varieties, one a dwarf, another with pendulous branches, and still another with deeply and curiously cut leaves. The fact that none of these have come into more general cultivation would indicate that they have not proven themselves superior to the type.

P. californica, as its name indicates, is a native of the Pacific slope, and is quite distinct from the preceding. The leaves are oblong, sharp-pointed, and on petioles. The general form is much the same, and the broad, round head, sometimes ten to twelve feet across, is densely covered with the characteristic foliage, the leaflets being oblong-lanceolate, sharp-pointed, and on slender petioles. The flowers are white or pale rose, with orange-colored anthers somewhat prominent. Unlike most others of this class they are highly fragrant. They appear in May, and crown the whole shrub with their upright spikes as they rise above the surface of green. In cultivation this variety usually grows from twelve to twenty feet high, but there are said to be specimens in its native habitat even much larger.

HEDYSARUM.

Hedysarum multijugum is an exotic, belonging to the *Leguminosæ*, which has recently come to us from Mongolia. It has not yet been thoroughly tested either in England or this country, but gives promise of being a valuable addition to the list of our hardy and ornamental shrubs. In the home of its adoption it seldom attains a height of more than five feet. It has slender branches, covered when young with minute, silky hairs which clothe alike the pinnate leaves and petioles, giving the whole bush a tinge of gray. The foliage suggests the tropics as the place of nativity, and the blossoms add force to the suggestion. These come forward in June, and are borne in racemes eight to twelve inches long, and keep in good form for two or three weeks. In color they are rosy-purple and very attractive. In the Northern States it may be well to afford the plant slight protection, especially in the early stages of its growth, but in the South this will be unnecessary. The species will certainly prove a valuable acquisition in that section.

TAMARIX.

THE hardy tamarisks are beautiful shrubs, and worthy of a more prominent place in ornamental planting than in this country has yet been accorded them. While the genus is quite large, only a few species are suited to ordinary garden cultivation. For planting by the seaside, and in especially exposed situations, they are of the utmost value, as their long, slender branches, readily yielding to the winds, are seldom broken. Not even the salt spray, so fatal to nearly all other plants, does them permanent harm, unless it be abundant and continuous. The tamarisks grow so rapidly as to be able to also serve a good purpose in such situations in affording protection to less hardy plants, and where their merits are appreciated they are sometimes grown in belts or masses for purposes of shelter. To make the most of them they must be severely cut back from year to year, as the tendency is to throw out long shoots that become bare stems, and as the foliage and flowers appear in most forms on the new growths only. They bear this clipping well, and are greatly improved in

Tamarisk.

their appearance by the operation. There is some confusion among specialists as to names both of the genus and the several species, but the following varieties will serve all practical purposes.

The African tamarisk, *T. africana*, is, perhaps, the earliest blooming species, the small flowers appearing in May or early in June in great profusion along the slender branches of the previous season's growth. They are bright pink and exceedingly beautiful, covering the whole bush and affording a marked contrast to the soft, feathery foliage. The time to cut in the branches is immediately after the flowering season is over, when the growth which follows will furnish blossom-buds for the next season. This tamarisk is known also as the *T. tetrandra*, and *T. parviflora*.

The French tamarisk, *T. gallica*, is a native not only of France, but also of Spain, Italy, and other Mediterranean countries. It grows to a height of eight or ten feet, with its very small, imbricated, feathery leaves in profusion, and the color contrasts are marked. The pale red or pink flowers are small but numerous, not distributed as in the *africana*, but appearing in clusters or catkins about an inch long at the end of the slender branches. The twigs are also conspicuous because of their purple or reddish-colored bark. The shrub can be shaped to a low, round, compact head, or be trained to almost any form that may be desired, and is sometimes grown to advantage on the side of a building or wall. In either case the lightness and grace of foliage and blossom are sure to attract attention and admiration.

Ornamental Shrubs.

The late-flowering tamarisk, commonly known as *T. indica*, unlike those already named, blooms on wood of the same season's growth, and so must be cut well back in late autumn or early spring, when the plant will put forth long, slender branches, six feet or more, which will produce a profusion of blossoms in August and September, affording a mass of color commingled with the fine, soft foliage that covers the spreading stems from end to end. "Nothing," says *Garden and Forest*, "can be more exquisitely graceful than the entire habit of this plant, and it is especially attractive in early morning when its branches droop under the weight of silvery dew." Neither the foliage nor the blossom differs in the early and late species to any appreciable degree, but the training of the plants must be quite unlike. If the early bloomers are cut back after the flowering season is over, the flower-buds for the next summer are all removed, and in this respect the season's growth will be a failure. But, if the branches of the *indica* are not cut sharply back at the close of the season, the plant will make another growth of six or eight feet, and the blossoms will appear only on the new wood, leaving a mass of bare stems below, which destroys the symmetry and attractiveness of the whole.

KERRIA—Globe Flower.

THERE is but one species of the kerria which is worthy of cultivation. It is a native of Japan and known as *K. japonica*, and popularly called Jews' Mallow, and is one of the early spring bloomers especially suitable for growing on a wall or fence. It is a deciduous,

Sophora.

erect, handsome, and hardy plant, and sends up numerous stems which will thrive in almost any good soil. The flowers are orange-yellow with five elliptical petals, obtuse and spreading. There are several varieties, double and single, the double being most in use, but the single having more beauty, and to be preferred for many situations. The foliage is bright glossy green, smooth above, and slightly hairy on the under side. The height is from three to six feet. There are one or two varieties having foliage variegated with white, and these cannot fail to be appreciated when better known.

SOPHORA.

THE sophoras include a number of small trees and shrubs, natives of eastern Asia and certain portions of North America, and possessed of much interest. They belong to the order *Leguminosæ*, and number not far from twenty species, with numerous varieties that are worthy of notice. The best-known member of the family is catalogued as *S. japonica*, though it is supposed to be a native of China, and to have been taken to Japan many centuries ago, carrying with it the popular name, pagoda tree, which is still applied to it in both countries. This name comes from the fact that it is largely planted in the vicinity of temples and public buildings.

KERRIA JAPONICA.

Ornamental Shrubs.

In a certain sense it is looked upon by the Eastern populace as an object of veneration and emblematic of sacred things. At its best it grows some thirty feet high, but in this country it seldom reaches that altitude. It has

WEEPING SOPHORA.

graceful foliage, and the leaves are large, green tinged with blue, and hold their color well. They are compound, with eleven to thirteen oblong-oval and pointed leaflets. The flowers are terminal, in long, open panicles and creamy-white in color. They appear in August and September. Wherever seen this sophora proves a handsome tree. *S. j. pendula* is a weeping variety, and one of the best weeping-trees known. It should be grafted on the parent stock at such a height as is desired and will then form a large compact head with the branches reaching to the base on every side. It is not often seen on our lawns because of the difficulty and cost of securing good specimens.

S. secundiflora is an American species with somewhat larger blossoms of rich violet color. These appear in June or early July. The leaves are more coriaceous and glossy than those of the *japonica*, but this form is not so hardy, and is not advised for northern planting. Its home appears to be in the extreme Southwest. It is abundant in Texas and Mexico, and is there a broad-leaved evergreen, growing to the height of six feet. *S. tetraptera* is a native of New Zealand, and has yellow flowers. It is a beautiful half-hardy shrub suitable for planting only in the Southern States unless afforded ample protection. It is deciduous, and grows to a height of twelve feet.

RHODOTYPOS.

THIS is a genus of only a single species, and is so closely allied to kerria as to be often confounded with it. By the botanists it is given the name *R. kerrioides* in consideration of the resemblance of both its flowers and foliage to *Kerria japonica*. It is a native of Japan, and was introduced into England as early as 1866, but has been practically unknown in American gardens until a much later date. The shrub rises ten to fifteen feet in height, and when grown on a wall has almost as wide a spread. It can be grown in this way, or kept in due limits as a somewhat straggling bush, as may be desired. The branches are numerous and quite twiggy, and are clothed with light green leaves, opposite, oblong-ovate, pointed, and soft or silky beneath. The flowers very much resemble large, single roses, and are borne in profusion at the ends of the branches, continuing in succession for a long period. Says *Garden and Forest:* "There is hardly a day from early June until frost comes when a well-grown specimen will not give a few sprays with single flowers at their extremities. The pure white blossom among the light green leaves is very attractive, and half a dozen of these sprays will help to add lightness and grace to a vase of the highly-colored flowers which usually prevail at this season." The plant is reasonably hardy, and can be depended upon both in the North and South.

PTELEA—Hop Tree.

THE ancients knew the elm by this name, which is now applied to a mere shrub because of a fancied resemblance in the fruit. It belongs to an entirely different order, *Rutaceæ*, and has little in common with the majestic elm. It rarely rises to more than eight feet. The pteleas make up a genus of six species, only one of which has a place among ornamental shrubs. *P. trifoliata*, known as the hop tree and also as swamp dogwood, is a much-branched shrub with alternate leaves usually in threes, pinnate, and with oblong or ovate leaflets. If the foliage is bruised or crushed it exhales a rather unpleasant odor. The flowers are greenish-yellow, and have a short

Laburnum.

calyx four- or five-parted and somewhat imbricated, with four or five petals of greater length and also imbricated. They appear in late May or early June. The variety known as the golden hop tree, *P. t. aurea*, is the same in all respects except that its foliage is a bright yellow, a color which it retains the entire summer, if given a sunny position. In this respect it is excellent for the shrubbery or border or as a single specimen on the lawn where a bit of contrasting color is desired. Planted with the *Prunus pissardia*, or mingled with scarlet-leaved shrubs and trees, the effect is fine.

LABURNUM.

THERE are three species of this genus of the order *Leguminosæ*, each having several varieties which are attractive and showy. They are small, upright, slender-growing trees, and can scarcely be planted amiss in the border or on the lawn. *L. vulgaris* is popularly known as golden-chain, getting the name from the shape and color of the blossoms, which are in long pendulous racemes of bright yellow, covered with soft pubescence, and hang among the leaves from April until June. They are succeeded by pods which continue long on the tree but are by no means unsightly. The leaves are compound with ovate-lanceolate leaflets, and the stems and branches are slightly bronzed. Among the desirable varieties are *L. v. aureum*, with golden foliage; *L. v. involutum*, with curled leaflets in the form of rings; *L. v. waterii*, with racemes longer than those of the others and more deeply colored; and *L. v. parkesii*, which has still more conspicuous blossoms. These grow to a height of twenty feet, as does *L. alpinum*, which is particularly distinguished as the Scotch laburnum and also has the name of golden-chain. It has much the same flower as the *vulgaris*, and is perhaps the hardiest member of the family. The laburnums are natives of Europe and Asia Minor.

CITRUS—Orange.

THE oranges are among the most interesting flower- and fruit-bearing shrubs and trees, but their free cultivation is limited to tropical and semi-tropical climates. They belong to the order *Rutaceæ*, and are found in the genus citrus, and are distinguished for their beautiful and fragrant blossoms, their attractive foliage, and the peculiar habit of bearing flowers and fruit at the same time. Most of the oranges are easily grown, and they come to maturity as early as do the peach or apricot. The citrus is thought to be the longest-lived tree in the world. It is a native of the warm valleys of the Himalayas, and of Persia, where specimens of great age are found, though there are no means of determining the years which they may have seen.

In the range of the Gulf States the oranges, as also the lemon, are at home, though not always to be depended upon in unusually severe winters. Among the varieties of interest for more general cultivation and for ornamental use is *C. trifoliata*, a native of Japan, which was introduced into English gardens some time since, and has been thoroughly tested as to its hardiness in that climate. The London *Garden* says that in some of the southern counties specimens have been growing well in the open ground,

Citrus—Orange.

and even bearing fruit in apparent perfection. Since its introduction into the United States it has proved entirely

TRIFOLIATE ORANGE.

hardy in the far South, and is now making an encouraging record farther north. In the Carolinas and Virginia it appears to thrive, and specimens have been known to have survived mild winters without unusual protection. As a pot or tub plant it requires no more care or attention than is given to *Hydrangea hortensia* and its varieties, to make its place sure in garden ornamentation. The foliage is good in form and color, and the creamy-white flowers have the true rich orange fragrance. The yellow fruit is about

OTAHEITE ORANGE.

one and a half inches in diameter, and hangs long on the bush. *C. trifoliata* bids fair to be of much service aside from its ornamental character. The Department of Agriculture is testing its merits as a basis for the more tender sorts, and by hybridizing, crossing, and budding it is hoped to secure a new class of orange trees and one better fitted than any heretofore known for varying climates. It is believed that by this means the orange-bearing line may be materially extended northward.

C. otaheite is a Chinese dwarf orange, and valuable for ornamental purposes only. It seldom grows more than three feet high, and is of bushy habit, much branched and

Citrus—Orange.

somewhat tortuously. It is a free bloomer and fruit bearer, and begins to bloom when not more than a foot high, and like a true orange bears flowers and green and ripe fruit at the same time. It is grown in the open in the South, but in the North must be treated as a houseplant. The flowers are, when well opened, fully an inch

FLOWERING BRANCH OF ORANGE.

across, and have creamy-white petals about a group of many erect stamens. They are deliciously fragrant. The fruit is small and of inferior quality, but it is edible and fully as good as some of the foreign oranges that are brought to our markets. It is seedless, well-colored, and remains long on the pert little tree whether in-doors or out, and with the fresh blossoms and the glossy leaves makes a beautiful plant. The *otaheite* endures the heat and gas of living-rooms well, and if kept moist and clean seldom fails to gratify the grower with a crop of miniature fruit. In the open it thrives in ordinarily good soil, but does better in partial shade than if exposed to the full power of the sun.

HALESIA—Snowdrop Tree.

THE halesias are coming to be recognized as among our best ornamental shrubs, and with reason. They belong to the order *Styracaceæ*, and, though the genus contains not more than six or eight species, they are widely distributed as to nativity over Europe, Asia, and America. There are three indigenous to the United States, one to China, and two or more to Japan. Wherever known they are highly appreciated among the plants of the class to which they belong. They are among the very floriferous shrubs, the pure white blossoms enveloping the whole plant and making it a conspicuous object. The leaves are medium in size, ovate-oblong, sharply pointed, and slightly dentate. They are borne on slender petioles. While the halesias are not overparticular as to soils and situations, they appear to enjoy shady positions and to have a preference for moist, sandy soils.

H. tetraptera is known as the four-winged snowdrop or silver-bell, and has its pure white flowers in fascicles containing nine or ten bell-shaped blossoms each. These come forth in early spring as soon as the foliage appears, and are borne on pedicels from the axils of the growth of the previous year. They are followed by a four-winged fruit. As the branches of the shrub are long and slender

HALESIA—SNOWDROP TREE.

and very numerous, the well-rounded head shows to the best advantage. The plant is a native of this country. It grows to a height of some twenty feet.

H. diptera is also of American origin, but grows only ten feet high. It has even larger blossoms and leaves than the *tetraptera*, and many prefer it for garden planting as it is known to be equally hardy. *H. hispida* is a native of Japan, and has flowers in more corymbose racemes than has either of the preceding. The fruit is covered with stiff hairs. It is not yet much grown in this country, and has no especial merits over our own halesias. None of the halesias can be depended upon to withstand the winters of the extreme North unless well protected.

MISCELLANEOUS.

IN portraying the characteristics of ornamental shrubs, it has become evident that reference might well be made to some of the smaller members of not a few genera among the large trees, not belonging to the class described. This has already been done to some extent, but there are yet others of the lower forms in use in horticulture with especial features that should be mentioned in order to more full and complete information concerning general gardening. This chapter is accordingly added as a further help to readers of this volume.

What are known as the Japanese maples have been described in detail, but there are several other small forms of almost or quite equal value in garden planting, which we here proceed to characterize : *Acer campestre*, the English or cork-bark maple, a native of central Europe, grows to a height of from fifteen to twenty feet, and is of stocky, roundish habit and handsome foliage. The bark is, as suggested by the popular name, thick, rough, and somewhat corky. *A. colchicum rubrum*, the red colchicum maple, is from Japan, ten to fifteen feet high, of good form, with bright crimson-colored foliage when first grown, and is a rare and beautiful variety, but not entirely hardy in New England. *A. wierii laciniatum*,

Ornamental Shrubs.

Wier's cut-leaf maple, is one of the most useful of the family, as it is a rapid and graceful grower, forming beautiful specimens in a short time. It has pendulous branches, with deeply cut foliage, and as it becomes quite large is coming to be planted as a street tree. The ash-leaved maple, or box-elder, also has the advantage of being a rapid grower, and has light green, yellowish bark. There are two varieties of this species, one of which has its foliage marked with yellow, and the other with white. The yellow form is esteemed the more hardy. *A. Pennsylvanicum*, or the striped-bark maple, is a native tree, with broad and effective foliage, and well worthy of planting in all ordinary collections. *A. schwedlerii* is distinguished by its bright crimson foliage in the early part of the season, later taking on a purplish green. In autumn it again becomes crimson, and contrasts finely with other foliage when planted in groups. *A. worleii* is a golden-leaved sycamore maple, the foliage being bright yellow in spring and changing to a duller shade as the season advances. It is one of the most effective of the whole group for garden planting. *A. tartaricum*, or Tartarian maple, is of a more shrubby growth, and of irregular, rounded form. The leaves are rather light-colored, and the bark smooth. *A. ginnala* is described as a miniature maple from Siberia, with deeply notched leaves which take on most gorgeous colors in autumn—orange, crimson, and dark purple or black. In speaking of the *Acer rubrum*, Mr. Samuel Parsons, Jr., in his work on *Landscape Gardening*, says: " The most brilliant effects are reached in the red or crimson tints. Scarlet is a color almost unknown to the normal

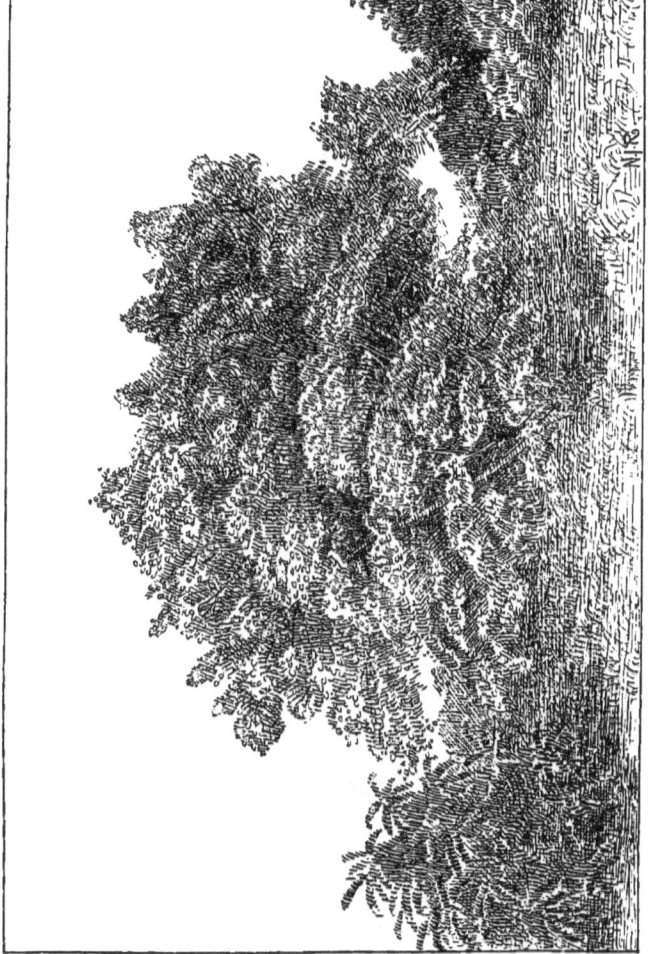

TARTARIAN MAPLE.

foliage of hardy plants. The most familiar example of this rich chord of color is found in the autumn tints of the swamp, or falsely named scarlet, maple, *Acer rubrum*. . . . The scarlet or red maple is the richest in autumnal color of all maples ; I was about to say of all trees. It seldom fails during any autumn to change more or less splendidly ; and therefore deserves to stand out a single flaming monument in the van of all autumnal color. There is something quite indescribable in the glow and intensity of tint often displayed by this maple. Is it ignorance or the want of seeing eyes that causes its lack of employment on the lawn? It is true the scarlet maple is slower growing than the sugar-maple, of less regular and pleasing outline, and certainly less beautiful and satisfactory at other seasons of the year. But in fall it simply reigns supreme."

As is well known, the catalpas flower in midsummer, and are as handsome as the horse-chestnut, the blossoms somewhat resembling those of that well-known species. One of the best forms of the smaller kinds is *C. bungei*, which comes from China, and grows only from three to five feet in height. Its foliage is large and glossy. It fails to be a first-class bloomer, but should not be

CATALPA BIGNONIOIDES.

Miscellaneous.

overlooked. *C. bignonioides* is a native of the United States, and is a showy, spreading, and irregular tree with heart-shaped leaves and pyramidal clusters nearly or quite a foot in length, and with white or purple fragrant blossoms. The golden catalpa is a variety of this species, differing from it chiefly in having leaves yellow in spring and early summer, afterwards becoming green. *C. kæmpferi* comes from Japan, produces yellowish-white flowers in June, and has smaller and somewhat distinct foliage. What is known as Teas's Japan hybrid is a low form with a spreading habit, having flowers with purple dots. The blossoms are fragrant and exceedingly abundant, and have the advantage of continuing for several weeks.

The beeches furnish also some peculiar and interesting forms. Among these is the well-known purple-leaved beech, which is probably the finest dark-leaved tree in cultivation. It grows large, is of symmetrical form, and, though quite out of the range of shrub life, is highly desirable on tree- and shrub-planted lawns. There is also a pendulous variety, established by grafting, with the same dark foliage and trailing branches. There are various forms of cut-leaf and fern-leaf beeches which must not be overlooked, though eventually they become too large to be classed with shrubs. What is known as the fern-leaf beech, *Fagus heterophylla*, is especially beautiful. It is believed that the first specimen brought into the country is still living, and stands in front of Redwood Library, on Bellevue Avenue, in Newport, where it is looked upon almost with veneration by the people of that city and those who make it their temporary home in summer.

Ornamental Shrubs.

It is a tree of compact and elegant habit, with its foliage finely cut and fern-like, and in spring is certainly one of the most charming specimens in field or garden. Every one in that city of villas makes it a point to secure one or more of these trees within his grounds. There are many other cut-leaved forms, but none so beautiful and delicate as

CUT-LEAVED BEECH.

this. The variety known as the weeping beech is too familiar to need description, but it must not be overlooked in making selections. *F. tricolor* is a variety, probably, of the purple beech, with a distinct border of rose color; but it is scarcely in general use, and probably for the reason that the variegation will not continue under the hot suns of summer. If planted at all, it should be in situations protected during the middle of the day.

Miscellaneous. 331

Salisburia adiantifolia is a remarkable tree, which was introduced some years since from Japan, growing at its maturity to a height of about forty feet. It is usually of slender form, and so suited to positions that are limited where a tree of that height is desired. It is known as the maidenhair tree, or gingko, having foliage resembling in form that of the adiantum fern, which is thick and glossy. In the cities of Europe it is becoming a favorite street tree, and is always handsome.

Cladrastis tinctoria, formerly known as *Virgilia lutea*, or yellow wood, is in all respects a most desirable tree of moderate growth, broadly rounded head, and compound foliage, of light green color, turning to yellow in autumn. The flowers are pea-shape, white, and fragrant, appearing in June in great profusion. It is not supposed to be sufficiently hardy for the colder portions of New England and the Northwest, but in southern Rhode Island and Connecticut it is found to withstand the climate in severest winters. More attention should be given to this beautiful tree than it has had in the past.

Of the birches there are also several interesting forms which cannot well be overlooked in garden planting. *Betula alba* is the well-known European white weeping birch, of rapid growth, with the bark of its stem and branches of a color most effective in winter, and a plant worthy of general attention. *B. alba aurea* is, perhaps, a more striking novelty. Its characteristics distinguishing it from the former are that the leaves in summer are of constant yellow, and associated with the purple and white form it becomes of great value. It is not much known

Ornamental Shrubs.

as yet in cultivation. *B. pendula laciniata* is the well-known weeping birch so largely in cultivation. It grows in slender form, having finely cut foliage on drooping branches. It is one of the best of the silvery-white forms for planting upon lawns, and should not be overlooked.

WEEPING BIRCH.

B. atro purpurea is a variety possessing the habit of the other birches, and distinguished from them by having purple foliage which is especially attractive in contrast with the white bark of the stems and branches. *B. nana*

Miscellaneous. 333

is, perhaps, the smallest member of the family known in cultivation. It is a bushy, shrubby tree, attaining about ten feet in height, and in every way attractive.

Many of the willows are also midway between trees and shrubs, and as such are of the utmost value in garden planting. They are especially to be recommended for winter effects. *Salix vitellina* is a small, shrubby form

WEEPING WILLOW.

having yellow bark, especially in winter, when showy effects are so much desired. There are other forms known as golden willows perhaps equally good, but none better than this. *S. pentandra*, or *laurifolia*, is one of the handsomest trees to be seen in any collection. The leaves are dark glossy green, and highly ornamental. It is also excellent for seashore planting, and withstands winds and cold much better than most plants. It can be grown in shrubby form if desired. *S. regalis* is worthy of large use for the contrasts furnished by its light or silvery foliage with other plants, and *S. rosmarinifolia*, or rosemary willow, has long, narrow, silky foliage, and is capable of being grown in a globular head by means of grafting.

INDEX.

A

Abelia, 92
Acacia, 44
Acer, 48, 326
Æsculus, 306
Ague Tree, 174
Alder, 236
Almond, flowering, 257
Alnus, 236
Althea, 302
Amelanchier, 176
Amorpha, 201
Andromeda, 225
Angelica Tree, 209
Apple, flowering, 104
Aralia, 299
Arrow-wood, 185
Azalea, 24

B

Baccharis, 224
Barberry, 110
Beach Plum, 261
Bechtel's Crab, 109
Beech, 329
Berberis, 110
Birch, 331
Black Alder, 143
Black Haw, 189
Black Wattle, 45
Box, 189
Box-elder, 326
Buckeye, 307
Buckthorn, 70

Buffalo Berry, 56
Burning Bush, 206
Bush Honeysuckle, 242
Buxus, 189

C

Calico Bush, 1
Calluna, 67
Calophaca, 294
Calycanthus, 12
Camellia, 272
Cape Jessamine, 119
Caragana, 264
Cassandra, 231
Catalpa, 328
Cercis, 288
Chaste Tree, 275
Cherry, flowering, 247
Chinese Crab, 107
Chinese Lilac, 217
Chionanthus, 239
Citrus, 318
Cladrastris, 331
Clethra, 9
Colutea, 72
Cornel, 276
Cornelian Cherry, 284
Cornus, 276
Corylopsis, 291
Corylus, 171
Crab, 104
Crape Myrtle, 174
Cratægus, 74
Currant, flowering, 155
Cydonia Japonica, 102

335

Index.

D
Daphne, 178
Desfontainea, 35
Desmodium, 210
Deutzia, 4
Diervilla, 150
Dirca, 35
Dogwood, 276
Dwarf Horse-chestnut, 306

E
Elæagnus, 266
Elder, 15
Erica, 67
Euonymus, 205
Exochorda, 14

F
False Indigo, 201
Fatsia, 302
Fern-leaf Beech, 329
Forsythia, 33
Fragrant Sumach, 195
Fringe Tree, 239

G
Gardenia, 119
Garland Flower, 180
Ghent Azaleas, 28
Gingko, 331
Globe Flower, 312
Golden Catalpa, 329
Golden Chain, 318
Golden Hop Tree, 317
Golden Willow, 333
Gordonia, 287
Great Laurel, 168
Groundsel Tree, 224
Guelder Rose, 185

H
Halesia, 322
Halimifolia, 224
Hamamelis, 292
Hardhack, 123
Hawthorn, 74

Hazel-nut Tree, 171
Heath, 67
Hedysarum, 309
Hercules Club, 299
Hibiscus, 302
High-bush Cranberry, 183
Hobble Bush, 182
Holly, 136
Hop Hornbeam, 209
Hop Tree, 316
Horse-chestnut, 306
Horse Sugar, 145
Hydrangea, 36
Hypericum, 60

I
Ilex, 136
Ironwood, 209
Itea, 274

J
Japanese Maples, 48
Japan Quince, 102
Jew's Mallow, 312
Judas Tree, 288
June Berry, 177

K
Kalmia, 1
Kerria, 312
Kœlreuteria, 192

L
Laburnum, 317
Lagerstrœmia, 174
Lead Plant, 202
Leatherwood, 35
Leucothoë, 233
Ligustrum, 57
Lilac, 211
Loblolly Bay, 288
Lonicera, 242

M
Macrostachya, 306
Magnolia, 81
Maidenhair Tree, 331

Index.

Mallow, 302
Maple, 48, 326
Meadow Sweet, 122
Mock Orange, 294
Morus, 99
Mountain Laurel, 1
Mulberry, 99
Myrica, 21

N

Naked Viburnum, 184
Ninebark, 121

O

Oleo Fragrans, 204
Orange, 318
Osmanthus, 202
Ostrya, 209

P

Pagoda Tree, 313
Paulownia, 18
Pavia, 306
Peach, flowering, 259
Pearl Bush, 14
Persian Lilac, 216
Philadelphus, 294
Phillyrea, 66
Pieris, 227
Pinxter, 28
Plum, flowering, 261
Prim, 57
Prinos, 143
Privet, 57
Prunus, 247
Ptelea, 316
Purple-leaved Beech, 329
Pyrus Japonica, 102
Pyrus Malus, 104

R

Red Bud, 288
Red Osier, 282
Rhamnus, 70
Rhododendron, 161

Rhodotypos, 315
Rhus, 194
Ribes, 155
Rosa Rugosa, 95
Rose Acacia, 46
Rosemary Willow, 334
Rose of Sharon, 303
Rowan Tree, 182

S

Salisburia, 331
Sambucus, 15
Sassafras, 172
Scarlet Maple, 328
Scotch Laburnum, 318
Service Berry, 177
Shad Berry, 177
Sheep Berry, 183
Shepherdia, 56
Siberian Crab, 108
Siberian Pea Tree, 264
Silver Bell, 322
Smoke Tree, 198
Snowball, 184
Snowdrop Tree, 322
Sophora, 313
Spindle Tree, 206
Spiræa, 120
Spurge Laurel, 181
Stagger Bush, 228
Staghorn Sumach, 195
Steeple Bush, 123
Stephandra Flexuosa, 71
St. John's-wort, 60
St. Peter's-wort, 126
Strawberry Tree, 205
Stuartia, 157
Styrax, 146
Sumach, 194
Swamp Dogwood, 316
Swamp Honeysuckle, 28
Swamp Rose, 303
Sweet Fern, 24
Sweet Leaf, 145
Sweet Pepper-bush, 9
Sweet Viburnum, 183
Symplocus, 144

Index.

Syringa, 211
Syringa (Philadelphus), 294

T

Tamarix, 310
Tartarian Maple, 326
Thorn, 74
Tree Box, 190

V

Venetian Sumach, 198
Viburnum, 181
Virgilia, 331
Vitex, 275

W

Waahoo, 206
Wax Myrtle, 22
Wayfaring Tree, 182
Weeping Beech, 330

Weeping Birch, 332
Weeping Sophora, 315
Weigela, 150
Weir's Cut-leaf Maple, 326
White Weeping Birch, 331
Wig Tree, 198
Wild Rosemary, 226
Willow, 333
Winterberry, 143
Witch Hazel, 292
Wythe Rod, 184

X

Xanothoceras, 134

Y

Yellow Wood, 331

Z

Zenobia, 232

BOOKS FOR THE COUNTRY

NATURE STUDIES IN BERKSHIRE
By JOHN COLEMAN ADAMS. With 16 illustrations in photogravure from original photographs by ARTHUR SCOTT. 8°, gilt top.

A collection of prose pictures of skies and woods and fields, intermingled with the reflections of a writer who is at once a philosopher and a poet, one who enjoys profoundly the beauties of the Berkshire Hills, and who possesses the art of enabling his reader to share in his enjoyment.

LANDSCAPE GARDENING
Notes and Suggestions on Lawns and Lawn-Planting, Laying out and Arrangement of Country Places, Large and Small Parks, Cemetery Plots, and Railway-Station Lawns; Deciduous and Evergreen Trees and Shrubs, The Hardy Border, Bedding Plants, Rockwork, etc. By SAMUEL PARSONS, Jr., Ex-Superintendent of Parks, New York City. With nearly 200 illustrations. Large 8°, $3.50.

"Mr. Parsons proves himself a master of his art as a landscape gardener, and this superb book should be studied by all who are concerned in the making of parks in other cities."—*Philadelphia Bulletin.*

LAWNS AND GARDENS
How to Beautify the Home Lot, the Pleasure Ground, and Garden. By N. JÖNSSON-ROSE, of the Department of Public Parks, New York City. With 172 plans and illustrations. Large 8°, gilt top, $3.50.

"Mr. Jönsson-Rose has prepared a treatise which will prove of genuine value to the large and increasing number of those who take a personal interest in their home grounds. It does not aim above the intelligence or æsthetic sense of the ordinary American citizen who has never given any thought to planting and to whom some of the profounder principles of garden-art make no convincing appeal."—*Garden and Forest.*

ORNAMENTAL SHRUBS
For Garden, Lawn, and Park Planting. With an Account of the Origin, Capabilities, and Adaptations of the Numerous Species and Varities, Native and Foreign, and Especially of the New and Rare Sorts, Suited to Cultivation in the United States. By LUCIUS D. DAVIS. With over 100 illustrations. 8°.

This volume is addressed to both scientific men, and that large class of persons who, though interested in plants, have no knowledge of Botany, and neither time nor inclination to acquire it. The phraseology is plain and the descriptions are easily comprehensible; yet the book contains material never before presented, relating to varieties of plants developed under cultivation.

THE LEAF COLLECTOR'S HANDBOOK AND HERBARIUM
An aid in the preservation and in the classification of specimen leaves of the trees of Northeastern America. By CHARLES S. NEWHALL. Illustrated. 8°, $2.00.

"The idea of the book is so good and so simple as to recommend itself at a glance to everybody who cares to know our trees or to make for any purpose a collection of their leaves."—*N. Y. Critic.*

THE WONDERS OF PLANT LIFE
By Mrs. S. B. HERRICK. Fully illustrated. 16°, $1.50.

The only thing aimed at is to give the more important types in a popular way, avoiding technicalities where ordinary language could be substituted, and, where it could not, giving clear explanations of the terms.

"A dainty volume . . . opens up a whole world of fascination . . . full of information."—*Boston Advertiser.*

G. P. PUTNAM'S SONS, 27 & 29 West 23d St., New York

BOOKS FOR THE COUNTRY

OUR INSECT FRIENDS AND FOES
How to Collect, Preserve and Study Them. By BELLE S. CRAGIN. With over 250 illustrations. 8°.

Miss Cragin sets forth the pleasure to be derived from a systematic study of the habits of insects, and gives many points which will be of practical value to the beginner. She gives comprehensive descriptions of all the more important species to be found in the United States, together with illustrations of the same.

AMONG THE MOTHS AND BUTTERFLIES
By JULIA P. BALLARD. Illustrated. 8°, $1.50.

"The book, which is handsomely illustrated, is designed for young readers, relating some of the most curious facts of natural history in a singularly pleasant and instructive manner."—*N. Y. Tribune*.

BIRD STUDIES
An account of the Land Birds of Eastern North America. By WILLIAM E. D. SCOTT. With 166 illustrations from original photographs. Quarto, leather back, gilt top, in a box, *net*, $5.00.

"A book of first class importance. . . . Mr. Scott has been a field naturalist for upwards of thirty years, and few persons have a more intimate acquaintance than he with bird life. His work will take high rank for scientific accuracy and we trust it may prove successful."—*London Speaker*.

WILD FLOWERS OF THE NORTHEASTERN STATES
Drawn and carefully described from life, without undue use of scientific nomenclature, by ELLEN MILLER and MARGARET C. WHITING. With 308 illustrations the size of life, and Frontispiece. New edition in smaller form. 8°, *net*, $3.00.

"The authors of this excellent work offer it, not in competition with scientific botanies, but with the hope that by their drawings and descriptions they may make it easy to become acquainted with the wild flowers of the northeastern portion of the United States. Anybody who can read English can use the work and make his identifications, and, in the case of some of the flowers, the drawings alone furnish all that is necessary. . . . The descriptions are as good of their kind as the drawings are of theirs."—*N. Y. Times*.

THE SHRUBS OF NORTHEASTERN AMERICA
By CHARLES S. NEWHALL. Fully illustrated. 8°, $1.75.

"This volume is beautifully printed on beautiful paper, and has a list of 116 illustrations calculated to explain the text. It has a mine of precious information, such as is seldom gathered within the covers of such a volume."—*Baltimore Farmer*.

THE VINES OF NORTHEASTERN AMERICA
By CHARLES S. NEWHALL. Fully illustrated. 8°, $1.75.

"The work is that of the true scientist, artistically presented in a popular form to an appreciative class of readers."—*The Churchman*.

THE TREES OF NORTHEASTERN AMERICA
By CHARLES S. NEWHALL. With illustrations made from tracings of the leaves of the various trees. 8°, $1.75.

"We believe this is the most complete and handsome volume of its kind, and on account of its completeness and the readiness with which it imparts information that everybody needs and few possess, it is invaluable."—*Binghamton Republican*.

G. P. PUTNAM'S SONS, 27 & 29 West 23d St., New York

www.ingramcontent.com/pod-product-compliance
Lightning Source LLC
Chambersburg PA
CBHW030314240426
43673CB00040B/1157